Cataclysmic Variable Stars
How and why they vary

Springer
London
Berlin
Heidelberg
New York
Barcelona
Hong Kong
Milan
Paris
Santa Clara
Singapore
Tokyo

Dr Coel Hellier
Lecturer in Astrophysics
Department of Physics
Keele University
Staffordshire

SPRINGER–PRAXIS BOOKS IN ASTRONOMY AND SPACE SCIENCES
SUBJECT *ADVISORY EDITOR*: John Mason B.Sc., Ph.D.

ISBN 1-85233-211-5 Springer-Verlag Berlin Heidelberg New York

British Library Cataloguing-in-Publication Data
Hellier, Coel
  Cataclysmic variable stars : how and why they vary. -
  (Springer–Praxis books in astronomy and space sciences)
  1. Cataclysmic variable stars
  I. Title
  523.8'446

ISBN 1-85233-211-5

Library of Congress Cataloging-in-Publication Data
Hellier, Coel
  Cataclysmic variable stars : how and why they vary / Coel Hellier.
    p.cm. – (Springer–Praxis books in astronomy and space sciences)
  Includes bibliographical references and indexes.
  ISBN 1-85233-211-5 (alk. paper)
  1. Cataclysmic variable stars. I. Title. II. Series.
  QB837.5 H45 2001
  523.8'466–dc21   00-051608

Apart from any fair dealing for the purposes of research or private study, or criticism or review, as permitted under the Copyright, Designs and Patents Act 1988, this publication may only be reproduced, stored or transmitted, in any form or by any means, with the prior permission in writing of the publishers, or in the case of reprographic reproduction in accordance with the terms of licences issued by the Copyright Licensing Agency. Enquiries concerning reproduction outside those terms should be sent to the publishers.

© Praxis Publishing Ltd, Chichester, UK, 2001
Printed by MPG Books Ltd, Bodmin, Cornwall, UK

The use of general descriptive names, registered names, trademarks, etc. in this publication does not imply, even in the absence of a specific statement, that such names are exempt from the relevant protective laws and regulations and therefore free for general use.

Cover design: Jim Wilkie

Printed on acid-free paper supplied by Precision Publishing Papers Ltd, UK

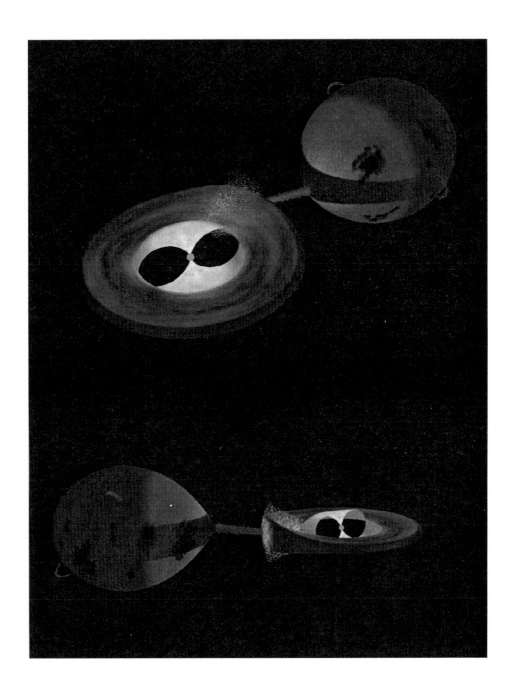

Coel Hellier

Cataclysmic Variable Stars

How and why they vary

Springer

Published in association with
Praxis Publishing
Chichester, UK

For Ewan and Robin

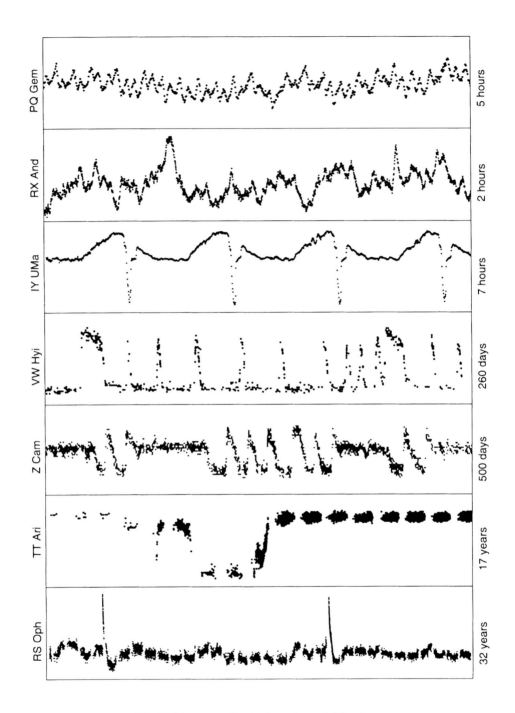

The lightcurves of cataclysmic variable stars

Contents

Preface	xiii

1 Observing cataclysmic variables   1
    1.1 Visual observations 2
        1.1.1 Finding the star 2
    1.2 Estimating the star's brightness 3
    1.3 CCD observations 7
        Box 1.1: Reducing a CCD frame 8
        1.3.1 Calculating the magnitudes 9
        1.3.2 Optimising the observations 11
        Box 1.2: Soft apertures 11
        1.3.3 Choosing a CCD camera 12
        Box 1.3: Image scale and field size 13

2 The orbital cycle   15
    2.1 The white dwarf 15
    2.2 The red dwarf 17
    2.3 The Roche geometry 17
        Box 2.1: Keplerian motion 18
        Box 2.2: NN Ser's component stars 19
        2.3.1 Ellipsoidal variations 21
    2.4 Mass transfer 21
        Box 2.3: The Roche geometry 22
        Box 2.4: Formulae for the Roche geometry 24
        2.4.1 The accretion disc 26
        2.4.2 The bright spot 27
        2.4.3 Grazing eclipses 29

3 Spectral characteristics   31
    3.1 The white dwarf spectrum 31
    3.2 The red dwarf spectrum 33
    3.3 The accretion disc spectrum 34
        3.3.1 Emission lines versus absorption lines 35

	3.4	Doppler shifts	37
		Box 3.1: Accretion disc temperatures	38
		Box 3.2: Eclipse mapping	39
		3.4.1 S-waves	40
		3.4.2 Double-peaked lines from the accretion disc	41
		Box 3.3: Doppler tomography	43
	3.5	Deriving masses and other parameters	44

4 The evolution of cataclysmic variables — 45
4.1	The origin of cataclysmic variables	45
4.2	Driving mass transfer	46
	4.2.1 Gravitational radiation	47
	4.2.2 Magnetic braking	47
	Box 4.1: The binary's response to mass transfer	48
4.3	The distribution of orbital periods	49
	Box 4.2: Estimating distances and mass-transfer rates	50
	4.3.1 The long-period cutoff	51
	4.3.2 The period gap	51
	4.3.3 The period minimum	52
	4.3.4 The ultimate fate	54
4.4	AM CVn stars	54

5 Discs and outbursts — 55
5.1	Dwarf nova outbursts	55
	Box 5.1: Osaki's argument for a disc instability	58
5.2	Viscosity in an accretion disc	59
	Box 5.2: Angular momentum exchange in an accretion disc	60
	5.2.1 Magnetic turbulence	62
5.3	The thermal instability	63
	5.3.1 Heating and cooling waves	65
	Box 5.3: The origin of the S-curve	65
	5.3.2 Outburst shapes	69
5.4	Novalike variables and Z Cam stars	72

6 Elliptical discs and superoutbursts — 75
6.1	Elliptical discs	75
6.2	Tidal torques and resonances	77
	Box 6.1: 'Beating' of two periods	77
	6.2.1 The effect of the orbital period	80
	Box 6.2: Mass ratios for resonance	80
	6.2.2 The superhump lightsource	82
6.3	The evolution of superhumps	82
	Box 6.3: The $O-C$ diagram	85
6.4	The supercycle	86
	6.4.1 Enhanced mass transfer?	87

6.5	ER UMa stars and WZ Sge stars	87
	6.5.1 EG Cnc and 'echo' outbursts	90
6.6	Permanent superhumps	91
6.7	Negative superhumps, or 'infrahumps'	92
	6.7.1 The problem of TV Col	93
6.8	Spiral shocks	93

7 Siphons, winds and streams   97
7.1	The boundary layer	97
	7.1.1 Siphons	97
7.2	Winds	99
	Box 7.1: The temperature of the boundary layer	100
	7.2.1 P Cygni profiles	101
	7.2.2 Winds in cataclysmic variables	102
7.3	The disc–stream impact	103
7.4	The SW Sex phenomenon	105
	7.4.1 A flared disc?	105
	7.4.2 Stream–disc overflow?	106
	7.4.3 Winds?	106
	7.4.4 Infrahumps?	107
	7.4.5 Relation to other novalikes	108

8 Magnetic cataclysmic variables I: AM Her stars   109
8.1	Magnetic accretion	109
8.2	The highest-field systems	110
8.3	The accretion stream in AM Her stars	111
8.4	AM Her X-ray lightcurves	113
	Box 8.1: Magnetic accretion	114
8.5	Cyclotron emission	116
	8.5.1 Polarisation	117
8.6	The accretion region	120
8.7	Asynchronous polars	123
	8.7.1 The origin of synchronous rotation	124

9 Magnetic cataclysmic variables II: intermediate polars   127
9.1	V2400 Oph: a discless intermediate polar	127
	9.1.1 The spin period of a discless accretor	128
	Box 9.1: Fourier analysis	130
	9.1.2 Accretion of diamagnetic blobs	132
9.2	Disc formation	132
	9.2.1 Diagnosing the accretion mode	133
9.3	Disc-fed accretion in an intermediate polar	135
	9.3.1 The accretion footprints	137
	9.3.2 Accretion in XY Ari	137
	9.3.3 XY Ari's response to a disc instability	138

x Contents

    9.4 Pulsations in a disc-fed accretor 141
        9.4.1 Optical pulsations at the spin period 142
        9.4.2 Double-peaked pulsations 143
        9.4.3 Optical pulsations at the beat period 144
        9.4.4 DQ Her 145
    9.5 Propellers 145
        Box 9.2: Multiple orbital sidebands in intermediate polars 146
        9.5.1 AE Aqr 147
        9.5.2 WZ Sge 148
    9.6 Evolution of magnetic cataclysmic variables 149
        Box 9.3: The accretion column in intermediate polars 149

10 Flickering and oscillations   151
    10.1 The location of flickering 152
    10.2 Quasi-periodic oscillations 153
    10.3 Dwarf-nova oscillations 154
        10.3.1 Weakly magnetic white dwarfs? 155
        10.3.2 Beat frequencies from the magnetospheric boundary? 156
    10.4 GK Per 157
    10.5 QPOs in AM Her stars 158
    10.6 Pulsating white dwarfs 158

11 The nova eruption   161
    11.1 Thermonuclear runaways 161
        Box 11.1: The reactions of a thermonuclear runaway 163
    11.2 The expanding nova shell 163
    11.3 The recurrence of nova eruptions 165
    11.4 Recurrent novae 167
        11.4.1 Relation to supernovae? 167
    11.5 Bursts in neutron-star systems 168
    11.6 Super-soft sources 169
        11.6.1 Thermal-timescale mass transfer 170

12 Secondary star variations   171
    12.1 VY Scl stars 171
    12.2 Observing star spots 174
    12.3 Solar-type cycles 175
    12.4 Short-lived flares: bursts of mass transfer? 176
    12.5 Long-term variations in mass-transfer rate 178
        12.5.1 Irradiation-induced mass-transfer cycles 179

13 Variations on the theme   181
    13.1 X-ray binaries 182
        13.1.1 Soft X-ray transients 183
        13.1.2 High-mass X-ray binaries 184
        13.1.3 The micro-quasars 185

13.2	Accretion outwith binaries	187
	13.2.1 Discs around young stars	187

A	Deriving the stellar masses	189
A.1	Measuring i and q	190
A.2	Measuring K_1	190
A.3	Measuring K_2	191
A.4	Measuring M_1 directly	193
A.5	Estimating M_2 from $P_{\rm orb}$	193

B	Note on units and symbols	195
C	Time conventions	196
D	Variable star nomenclature	197
E	Variable star organisations	197
F	List of cataclysmic variables	198

Bibliography	201
Object index	207
Index	209

The colour frontispiece shows a computer graphic of a magnetic cataclysmic variable, created by Andrew Beardmore

Preface

The routine was now familiar. Being summer there was time after dinner for a game of snooker before driving up to the telescopes. Today, however, a band of thunderstorms had passed through and the clouds had not yet cleared — time for a second game. Afterwards, stepping outside again showed the clouds rapidly thinning to reveal the pristine magnificence of the southern sky as seen from the African semi-desert. Opening up the dome I pointed the 30-inch telescope at a star I had been monitoring for a week. The computer recorded the data: 10 000 photons in a second! All week the star had slowly varied between 1000 and 2000 photons per second, so surely I had, in haste, pointed the telescope at the wrong star. But checking the field confirmed that it was the correct star, brighter than it had ever been seen before! That night I watched as it gradually faded, and by dawn had returned to its normal brightness.

The star was a cataclysmic variable, one of a class of star that vary in brightness in a prolific number of ways. They are thus of great interest both to professional astronomers and to the many amateurs who monitor their light curves with small telescopes. To date, though, accounts of cataclysmic variables have tended to be brief — two or three pages in books with a wider remit — or highly detailed and technical tomes written for those with a PhD in the subject.

This book aims for the middle ground, explaining cataclysmic variables to the amateur astronomer, the undergraduate, those commencing a PhD, or anyone wanting an introductory account of the most variable stars in the sky.

To assist the different readers, the book can be read at several levels. The core text describes cataclysmic variables in a self-contained and non-mathematical way. Additional 'boxes' (set in a sans-serif font) give further details and physical equations, pitched at a level suitable for an undergraduate. I would advise those new to the topic to skip the boxes on a first reading, and return to them as interest or need dictates. Footnotes are also used to keep a detail or an equation out of the main text, while superscripted numbers refer to the bibliography, enabling the book to serve as an introduction to the professional literature.

One difficulty is that whereas undergraduate teaching uses SI units almost exclusively, many astronomers retain the older cgs units, while natural units such as 'solar masses' abound. I have addressed this by often quoting values in more than one form, and providing a note on conversion in the appendices. I have also tried to introduce and explain the technical terms as they arise, assuming only those —

such as stellar magnitude — that will be familiar to all astronomers.

Cataclysmic variable research is ongoing, and in places I emphasise that the explanations are incomplete and only partially understood. I have tried to present a coherent picture of the state of play, using my judgement on some issues of active debate, but future developments will no doubt require that this picture be revised.

I gratefully acknowledge those who have commented on drafts of this book, thereby much improving the end result. These are Koji Mukai, Janet Wood, Tim Naylor, Andrew Beardmore and the copy editor, Bob Marriott.

I am particularly grateful to Janet Mattei, of the American Association of Variable Star Observers, for permission to reproduce lightcurves and finding charts from their database. Likewise, Frank Bateson, of the Royal Astronomical Society of New Zealand, supplied lightcurves of southern cataclysmic variables.

In addition, the following are thanked for their assistance in compiling the illustrations: Andrew Beardmore, Klaus Beuermann, Albert Bruch, David Buckley, Sean Harmer, Jens Kube, Joe Patterson, Steve Potter, Dan Rolfe, Alan Smale, Barry Smalley, Henk Spruit, Danny Steeghs, Martin Still, Tod Strohmayer, Hans-Christoph Thomas, Pete Wheatley, Janet Wood, Graham Wynn, and Liza van Zyl.

Lastly, many of the illustrations were originally published in the following journals (as documented in the bibliography), and are reproduced with permission from the editors and copyright holders. *Monthly Notices of the Royal Astronomical Society* is published by Blackwell Science, with the copyright held by the Royal Astronomical Society. The *Astrophysical Journal* and *Astronomical Journal* are published by the University of Chicago Press, with the copyright held by the American Astronomical Society. *Astronomy and Astrophysics* is published by Springer–Verlag with the copyright held by the European Southern Observatory. The *Publications of the Astronomical Society of the Pacific* is copyright to the Astronomical Society of the Pacific, while the *Publications of the Astronomical Society of Japan* is copyright to the Astronomical Society of Japan.

Coel Hellier
Keele, October 2000

Chapter 1

Observing cataclysmic variables

Most stars are boring: stare at them and they would change incomprehensibly slowly, evolving only over hundreds of millions of human lifetimes. Perhaps only one in a thousand varies perceptibly in the time a human would care to watch it.

One type of star, however, a cataclysmic variable, can be relied upon to vary continuously and unpredictably, never exactly repeating itself. Observe a cataclysmic variable for half an hour and it may fade by a factor of ten, only to recover 10 minutes later. The next night it might be a hundred times brighter. Or it might have dropped beneath the threshold of visibility. Some exhibit a rhythmic pulse, brightening and fading every fifteen minutes. And all the while they will be flickering irregularly.

Amateur and professional astronomers alike are attracted by such restless stars. We have been recording the lightcurves of cataclysmic variables for more than a century, long before we had an inkling of what they were. More than a hundred are visible by eye with a small telescope — several dozen are monitored nightly by amateurs worldwide — and more than five hundred can be detected with the addition of a CCD camera. They are the most actively observed class of variable star amongst amateurs, a mainstay of project work amongst undergraduates studying astronomy, and the frequent target of the professionals' latest satellites and observatories.

In the four decades since Robert Kraft and colleagues established that cataclysmic variables are pairs of stars, orbiting so closely that stellar material flows between them, we have developed a good understanding of the dozen different types of variability they display.[1] That is the theme of this book.

In presenting an accessible account of cataclysmic variables, one aim is to inspire the reader to participate in one of the few areas of modern science to which the part-timer can make a contribution. The record of amateur observations of cataclysmics is not merely valuable, but, as will become apparent, is an essential underpinning of our understanding. There are too many cataclysmics for the professionals to monitor continuously, but there are many more amateurs. Thus when a cataclysmic variable does something unusual — and they do — it is the amateur who will spot it. The rest of this chapter explains how to make such observations. The reader who is less interested in the observational techniques, though, could skip straight to Chapter 2.

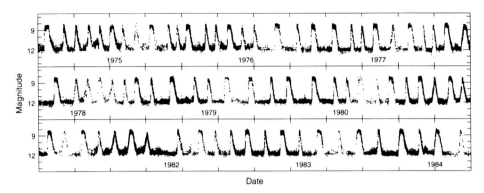

Fig. 1.1: Ten years in the life of SS Cyg, compiled by the American Association of Variable Star Observers (AAVSO). Tickmarks are at 100-day intervals.

1.1 VISUAL OBSERVATIONS

Point a pair of binoculars at the cataclysmic variable labelled 'SS' in the constellation Cygnus and you will see it on perhaps one out of four nights; the rest of the time it will be too faint. With a small telescope you will always see it, and can watch it vary between 8th and 12th magnitude in a semi-regular cycle lasting 50 days. Since 1896 amateur astronomers have made more than 200 000 estimates of SS Cygni's brightness, of which a portion is shown in Fig. 1.1.

Such small-telescope estimates are the foundation of our understanding of cataclysmic variables. Even in these days of CCD cameras they have not been superseded: a skilled visual observer can record the brightness of variable stars just as efficiently as one with a CCD camera, and it is important to continue visual observations for consistency, since human eyeballs and CCD cameras see colours differently.

The most common amateur telescopes — Newtonian or Schmidt-Cassegrain reflectors with apertures of 6 to 8 inches — are fine for such work, being able to detect a score of the brighter cataclysmics, at least at the brighter parts of their cycles. Obviously, though, with larger telescopes more targets become accessible. Of equal importance is the observing site — the darker the better — and clear skies! Whether the telescope is equipped with an equatorial drive is less important, since a low-power eyepiece is normally best for variable star estimates, and the Earth's rotation can be corrected for by using slow-motion controls.

1.1.1 Finding the star

The first task, and probably the hardest, is finding the target star. The time-honoured technique is called star-hopping. You will need a good sky atlas, showing stars down to magnitude 7–8, and a detailed star chart of the field around the variable (these can be obtained from the variable star organisations, see Appendix E). Using the star's coordinates, plot the position of the variable on the atlas and look for nearby bright stars that are part of a recognisable constellation. The idea is to point the telescope's finder at the nearest bright star, use it to identify fainter stars nearer the variable, and keep moving towards the variable, hopping from star to star, matching the stars to the atlas. At some point move over to the detailed star chart and identify the variable itself. Helpfully, the variable star charts are available in a range of scales, and are often plotted in reverse to match the view through a telescope.

If you have not practised this technique, or are not familiar with the major constellations, it is advisable not to look for a cataclysmic variable at first, since they are relatively faint stars and so are hard to find. Instead, use a book such as David Levy's *Observing Variable Stars: a Guide for the Beginner,* which describes the constellations and gives charts for much brighter variables. Algol, varying from 2nd to 3rd magnitude every 69 hours, is the classic variable star for the novice.

Here are some hints for the less experienced. Use a low-power eyepiece to see more stars simultaneously so that the patterns are easier to recognise. Determine how the scale of the charts relates to the view through the eyepiece.* Finding a variable star for the first time is hard; finding it again subsequently when the patterns are familiar is much easier, so have patience the first time. I remember a week-long observing run at an observatory where, with unfamiliar equipment, it took me an hour to find the target star on the first night. By the end of the week I could find it within a minute.

Of course if you have a modern computerised telescope you could by-pass all the fun by typing in the star's coordinates and watching the correct field pop up effortlessly (or not, if you haven't polar-aligned it properly!).

1.2 ESTIMATING THE STAR'S BRIGHTNESS

If we humans were asked to estimate the brightness of an isolated light on a quantitative scale we would do it very poorly. If instead, though, the task is to discern which of two lights is the brighter, we can do it to a high accuracy. That is the principle exploited by variable star observers. The chart for the variable is labelled with the magnitudes of many of the surrounding stars (it is usual to quote magnitudes to a tenth, omitting the decimal point in case it is confused with a star, so that 129 means magnitude 12.9). Fig. 1.2 shows the chart for SS Cyg, while for those south of the equator Fig. 1.3 shows that for VW Hyi. Note that variable stars are labelled by their abbreviated constellation and a one- or two-letter prefix, or, when all letter prefixes have been used, V (for variable) followed by a catalogue number. Thus two variables in Hercules are AM Her and V533 Her. For the full (and idiosyncratic) details see Appendix D.

---

*Since the right ascension coordinate is based on 24 hours, one minute of right ascension is the distance a star will trail in one minute as viewed with a stationary telescope (with the drive turned off). The declination coordinate, though, is based on 360° and so minutes and seconds of arc (right ascension) are not the same as arcminutes and arcseconds (declination). For stars near the equator, one minute of right ascension is equivalent to 15 arcminutes of declination; thus a star near the equator will trail 1 arcmin in 4 secs.

Since few stars are variables the surrounding stars can be used as constant comparisons. Thus the observer's task is to judge which comparison star is closest in brightness to the variable, but just a little dimmer, and which is closest but just a little brighter. The variable's magnitude is then between the magnitudes of these two. Furthermore, the observer can estimate that the variable is, say, one third of the way between the brighter and the dimmer comparison stars. Thus if these two were at magnitude 11.4 and 12.0 the variable would be estimated as 11.6. Given suitable comparison stars a skilled observer can judge a variable to 0.1 magnitudes, or an accuracy of about 10%.

Having outlined the principle, here are some details and hints.

- Record the time at least to the nearest few minutes. Make sure you record the time system used (see Appendix C for details of time systems) and other information such as the telescope and chart.

- If you cannot see the variable, look for the faintest star you can see and record the variable as being fainter. Such 'upper limits' can be valuable. For instance, if a variable is easily visible one night, a null result the previous night would prove that it had brightened over those 24 hours, and had not been bright for a week.

- If your estimate is less reliable than usual, perhaps because of cloud or moonlight, record this. The usual convention is to add a colon after the magnitude, giving 12.0: for example.

- The variable star organisations produce standardised record sheets for observations, and will be pleased to add your observations to their database.

- Move the variable and comparison stars around in the eyepiece. If the variable is always centred and the comparisons off-centre, the greater optical quality in the centre of the field may produce a systematic bias.

- Being properly dark adapted will enable you to see to fainter limits and will produce consistent results. Dark adaption is not only the widening of the pupil of the eye. More subtle changes include a reduction in the number of photons that the photocells in the retina need to accumulate before a signal is passed to the brain, and also an increase in the length of time over which these events can accumulate. A photocell in a fully dark-adapted eye (which needs 30 mins) can trigger on receiving only 7–8 photons every ~ 20 secs; in sunlight it triggers only on receiving 100 photons in 0.01 secs. Red light destroys dark adaption much less than white light, so use a weak, red flashlight (sparingly).

- For consistent results try to standardise your method. For instance, estimate magnitudes after having looked away from the field. Do not stare at the stars since light can accumulate, causing them to look brighter (see last item). Since this affects different colours differently, it can cause bias if the stars in the field have different colours.

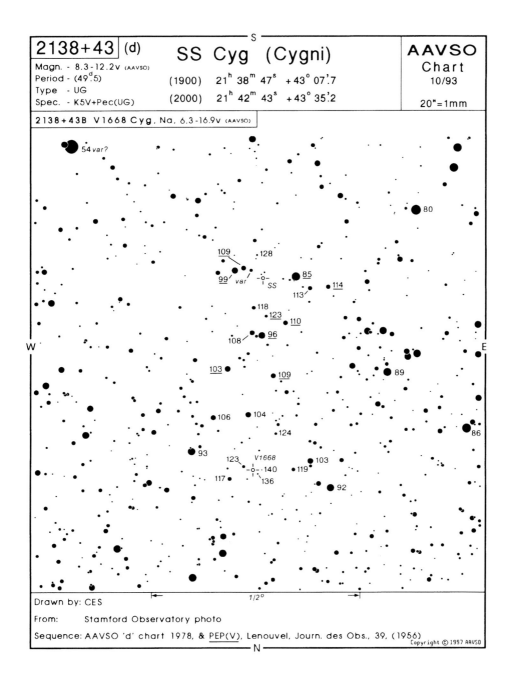

Fig. 1.2: The chart for SS Cyg published by the AAVSO. SS Cyg is labeled *SS* while another variable, V1668 Cyg, lies in the same field. (Copyright AAVSO, reproduced with permission.)

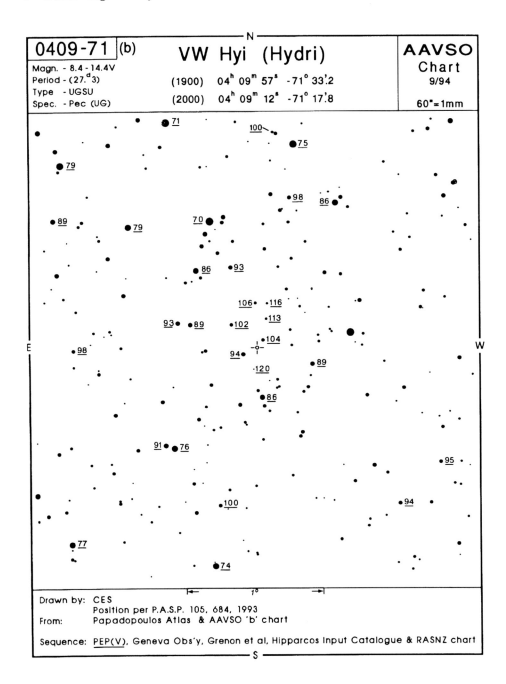

Fig. 1.3: The AAVSO chart for the bright southern cataclysmic variable VW Hyi (cross at center). VW Hyi varies from 8th to 14th magnitude every 30 days. (Copyright AAVSO, reproduced with permission.)

- Moonlight will brighten the sky background, making it harder to see faint stars. If the variable and the comparison stars have different colours the addition of blue moonlight can also cause bias. Thus it is tempting to avoid moonlit nights. But suppose the variable does something unusual near full Moon, will anyone be watching? Observations at such times (and also late or early in the season when the star is only observable in twilight) can be more valuable than those when the star is optimally located. There is a seasonal pattern in the density of observations of SS Cyg in Fig. 1.1.

- To see fainter stars try averted vision. This is a technique where one *looks* at one thing, but *pays attention* to something else (it can be tricky at first). The point is that the photocells in the centre of the retina, the fovea, are excellent at discerning detail, but are not very sensitive. Photocells off-centre are more sensitive, but are much worse at discerning detail. This difference is not normally noticed since rapid eye movements point the fovea at whatever catches the attention, but it becomes obvious if you try reading text at which you are not directly looking. If you use averted vision, ensure that you view the comparison stars with the same part of the retina.

- Although it is unlikely to be a problem with cataclysmics, avoid judging a variable with a telescope in which it appears very bright, since glare reduces the eye's ability to discern subtle differences. I learnt this looking at Jupiter with a 30-inch telescope during the week when it was bombarded by the comet Shoemaker–Levy: in twilight, with the sky the same brightness as the planet, incredible detail was visible; an hour later the detail was lost in the glare of the planet against a dark sky.

- When making an estimate, put out of your mind any previous estimates or expectations. Human judgement is notoriously unreliable when one thinks one knows the correct answer.

- The optimum frequency of observation depends on the star's behaviour. An observation every clear night will record the outburst properties (see Chapter 5), but if the star is changing rapidly, say on the rise to outburst, estimates every 30 mins are worthwhile. Some visual observers have detected the much more subtle 'superhumps' (Chapter 6) by making estimates every ten minutes. When attempting this it is vital to remember the previous point.

1.3 CCD OBSERVATIONS

Charge-coupled device cameras (CCDs) for small telescopes are now commonly available. CCD chips are wafers of semiconducting silicon divided into little squares called pixels. Photons hitting the device dislodge electrons from the semiconductor, creating pools of electric charge in each pixel. After allowing the charges to build up over a set 'exposure' time, the controlling electronics 'reads out' the device by shunting the charge from pixel to pixel and out of the side of the device, where it

Box 1.1: Reducing a CCD frame

One disadvantage of CCDs is that using them is more convoluted than making a visual estimate. Outlined below are the usual processing steps, although the details of the procedure will depend on the particular CCD camera and the software supplied with it.

SUBTRACTING THE BIAS
The process of reading out a CCD is not exact, but involves an uncertainty of a few counts, called readout noise. Because of this a CCD is usually set up so that the controller adds a set number of counts to each pixel, called the bias level. Thus the recorded counts will fluctuate about the bias level. (If there were no bias level, the counts would attempt to fluctuate about zero, but since the controller does not record negative counts a systematic error would result.) The bias can be found by simply reading out the CCD without making an exposure; the 'bias frame' should then be subtracted from all data frames.

SUBTRACTING A DARK FRAME
The thermal energy of molecules in the CCD chip can dislodge electrons, which accumulate over a CCD exposure and will be recorded as counts. To minimise this effect CCDs are cooled, often using liquid nitrogen (in professional observatories) or with a Peltier cooler operating at around $-30°$ C. To correct for the thermal electrons, called dark current, take a CCD frame as for the data frames, but without exposing the CCD to light. Subtract this frame from the data. Since the dark frame will also contain the bias level, subtracting both bias and dark current can be done in one step. Obviously the dark frame needs to be exposed for the same time as the data frames, and with the CCD at the same temperature. For greater accuracy record a set of about ten dark frames and combine them into an average dark frame to reduce statistical fluctuations (taking the 'median' of a pixel's value in each frame often works better than taking the mean, a process called 'median stacking').

FLAT-FIELDING
A CCD's pixels will not all have the same sensitivity to light. Thus if the variable lands on a pixel that records 66% of the incident photons, and the comparison on a pixel recording 63% of the photons, there would be a systematic error. To correct for this one uses a 'flat-field' frame, which is a frame exposed to a uniform light source such as the twilight sky (expose to at least 10 000 counts/pixel, and again a median stack of several frames increases the accuracy further). After subtracting the bias and dark current from this frame, divide the value of each pixel by the average of all the pixels in the frame. This produces a 'normalised' frame in which more sensitive pixels have values slightly above one and less sensitive pixels values below one. By dividing the (dark- and bias-subtracted) data frames by the normalised flat-field you effectively adjust every pixel to the same sensitivity.

COSMIC RAYS
Energetic particles caused by 'cosmic rays' striking the Earth's atmosphere can hit the CCD, causing trails of excess counts. If these affect the variable or comparison values, the data from that frame are best discarded.

Sec 1.3   **CCD observations**   9

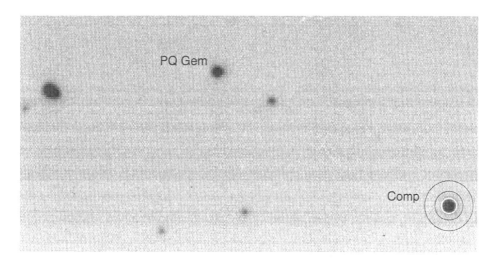

Fig. 1.4: A CCD image of PQ Gem. Only part of the chip has been read out, sufficient to record two brighter comparison stars, although one of these is a blend. The white aperture around the comparison was used in adding up the star counts. The annulus defined by the dark rings was used to estimate the sky contribution.

is recorded electronically and fed into a computer. The result is a two-dimensional image of the accumulated light seen by each pixel in the chip.

The advantages of a CCD are its greater sensitivity than even the dark-adapted eye and its ability to accumulate light over longer times. A CCD will typically see ~ 3–4 mags fainter than a human with the same telescope. Furthermore, the data are recorded digitally, eliminating human subjectivity, so it is easy to achieve $\sim 1\%$ accuracy, compared to 10% with an eyeball.

1.3.1 Calculating the magnitudes

The basic technique with a CCD is the same as with visual observations: the comparison of the variable's brightness with that of other stars recorded on the same image. A sample CCD frame, containing a target star and several field stars, is shown in Fig. 1.4 (the standard processing steps discussed in Box 1.1 have been applied).

The procedure is to add up the recorded counts from each star and adjust for the counts due to the sky background. The best way of doing this has been discussed at length in research papers, but the following outlines a reasonably simple method that works well and explains the principles (most CCD cameras are supplied with software that does something similar).

First find the coordinates of the centroids of the stars in which you are interested (place a small, square aperture about your star; average the x coordinates of each pixel in the aperture, but weight each pixel by multiplying by the number of counts it contains; then do the same for the y coordinates to obtain the x, y coordinates of the centroid). Now place a circular aperture about each centroid (we will discuss what radius it should have in a moment). Adding up the counts inside the circle gives the counts from that star, say Star A. Now place an annulus outside the circle and find the median value inside the annulus (we use a median, not a mean, in case the annulus contains faint stars). This median, multiplied by the number of pixels in the circle, is an estimate of the number of sky counts amongst those attributed to Star A, so subtract this number from Star A's total. Repeat the process for a comparison star (choose the brightest one in the field for greatest accuracy), ensuring that you use the same radius aperture.

Finally, divide the counts in the target by those of the comparison to obtain the variable's relative intensity. If the weather were perfect, the atmospheric turbulence constant and the altitude of the star (and hence the amount of atmosphere looked through) also constant, one would not need to take the ratio with another star and could just use the count rate of the variable. However, these conditions are never constant, and another great benefit of CCD photometry is that the field covers such a small area of sky that such factors affect all stars equally, to a very good approximation. Thus the observed brightness relative to other stars on the frame is far more robust than the raw count rate. If one uses the same comparison star for a set of CCD frames, the relative intensity is all that is needed to record the variable's lightcurve.

What radius should be used for the aperture round each star? That is a compromise. The limitations of the optics and the turbulence of the atmosphere ('seeing') blur the star's image into a profile with extended wings. A large aperture includes more of the star's light, but also more background sky counts, which reduces the accuracy of the measurement. A small aperture would minimise sky counts, but would exclude more of the starlight. Errors are also introduced if so few pixels are involved that pixel fractions become very important. The optimum will depend on the relative brightness of the sky and target star, so ideally one should experiment with different sizes to see which results in the minimum scatter. However, judging by eye the minimum radius which includes almost all the star's light will not be far wrong.

How can we use the results if we are not including the full wings of the stellar profile? Because the profile will be the same for the variable and the comparisons, so that the fraction of the light contained within a set radius will be the same for each. Thus the relative intensity of two stars will not depend on the radius chosen (which affects only the noise level of the measurement). It is, though, vital to use the same radius aperture for both stars, even if one is much brighter than the other.

Finally, if there is a third star in the field, of similar brightness to the variable, compute its ratio with respect to the comparison star. The scatter in this ratio over a set of CCD frames gives an estimate of the observational error in the measurements. It will also reveal whether you have inadvertently used a comparison star that is also variable! Fig. 1.5 shows a lightcurve of PQ Gem and a 'check' lightcurve, computed in this way.

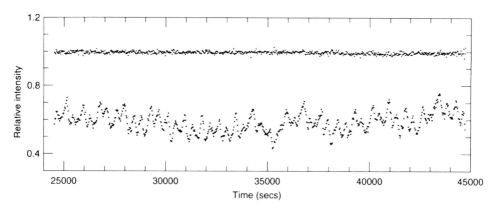

Fig. 1.5: A lightcurve of PQ Gem built up from 722 CCD frames, one of which is shown in Fig. 1.4. Each was exposed for 20 secs through an I-band filter on a 0.8-m telescope. The upper curve is the ratio of two comparison stars, used as a check. PQ Gem exhibits a 417-sec pulsation and a lot of 'flickering'.

1.3.2 Optimising the observations

EXPOSURE TIMES

The optimum exposure time depends on several factors. First, what timescale of variability are you trying to record? If the aim is to observe an eclipse lasting three minutes, you need to take a series of much shorter exposures, each one lasting no more than ~ 20 secs. Second, how faint are your stars? You need to expose for long enough to record enough counts for the accuracy required. Third, how good is your telescope tracking? You will not achieve more accurate results by longer exposures if the stellar profiles turn into wiggly star trails. Fourth, the CCD electronics can record only up to a set number of counts (typically 64 000). You need to ensure that neither the target nor comparison stars exceed this. A fifth consideration is the dead-time, the amount of time the CCD spends reading out between exposures.

Box 1.2: Soft apertures

When adding up the counts inside a circular aperture of radius r, how does one deal with fractions of a pixel? The simplest scheme is to ignore the pixel entirely if its centroid is further than r from the aperture centre, and include the full pixel if the centroid is less than r from the centre. However, since star images will fall differently with respect to pixel boundaries, this can amount to treating the variable and the comparison differently. The effect will be greatest with small apertures (minimising the sky contribution) or with large pixels comparable in size to the stellar profile. The countermeasure is to use a 'soft aperture'. Thus, if the pixel centroid is d from the centre of the aperture and both r and d are in units of pixels, ignore any pixel with $d > r + 0.5$, give full weight to any pixel with $d < r - 0.5$, and weight any pixel in-between by $r + 0.5 - d$.

If the dead-time is a large fraction of the exposure time, the observations will be inefficient. The best exposure time will thus be a compromise amongst many factors.

WINDOWING AND BINNING

A common way of reducing the dead-time is to 'window' the chip, which means reading out only part of it. However, you should include at least one and preferably two bright comparison stars (using the brightest available stars maximises the photon statistics), so the optimum windowing depends on the star field. 'Binning' means combining two or more adjacent pixels into one during readout, again to increase the speed. This is acceptable provided that the binned pixels are not too large (see Section 1.3.3).

SELECTING A FILTER

Your CCD may be equipped with filters to record light in a set of standard colours, normally ultraviolet, blue, visual, red and infrared, abbreviated to *UBVRI*. The disadvantage of using a filter is that recording only one colour yields fewer counts and so less accuracy. The advantage is that the results can be combined with those obtained by others, if they are also using standard colours. If you are looking for a rapid, low-amplitude pulsation, you might omit the filter for maximum sensitivity. However, if you are helping to build up the lightcurve of a star over months, in collaboration with other observers, you would all use the same filter.

For many purposes, particularly if not using a filter, relative photometry against the same comparison star is all that is needed. If, though, the magnitude of the comparison is known, the relative count rate can be translated into a magnitude. A CCD with a *V* filter has approximately the same sensitivity to colour as the eye, so the charts for visual observers can be used. However, as this is not exact you should always record the equipment and comparisons used, and report these with the results.

FOCUSING AND TIMING

In a properly focused image one can use small extraction apertures, reducing the amount of sky included, to obtain better results. It is thus worth checking the focus — which can be done by minimising the size of a star image — before taking a long series of CCD frames. Note, also, that such frames are of little value without accurate timing. Most PCs are not equipped with accurate clocks, so it may be necessary to set the clock each night, aiming for an accuracy of at worst a quarter of the shortest exposure time. The timings may also need to be adjusted for the difference between civil time and dynamical or heliocentric time (see Appendix C).

1.3.3 Choosing a CCD camera

Of the several makes of CCD available on the amateur market, which is the best for observing cataclysmics? Again the answer involves a compromise between competing factors, with the following issues being important.

PIXEL SIZE, FORMAT AND SKY COVERAGE

The area of sky covered by one pixel depends on the physical size of the CCD pixel (typically 10–20 μm) and the focal length of the telescope. The area should be small enough to sample the stellar profile adequately. A stellar profile is commonly described by its width when the intensity has dropped to half the peak value (the 'full width half maximum' or FWHM) and, ideally, each pixel should be no more than half the FWHM. The FWHM will depend on the quality of the optics, the atmospheric seeing, the precision of the focus, and the telescope tracking during the exposure. With amateur equipment it is likely to be 2–3 arcsec, so the ideal pixel size is about 1 arcsec. Larger pixels would force the use of larger apertures, including more sky. Smaller pixels would sample the image better, but only at the cost of decreasing the field size, reducing the chance of including bright comparison stars. One should aim for a field size of at least 4 arcmin square. Large-format chips, containing more pixels, can give both good sampling and good sky coverage. The main disadvantage is their greater cost; they also take longer to readout but this can be offset by windowing and binning as appropriate.

QUANTUM EFFICIENCY, LINEARITY AND GAIN

The quantum efficiency (QE) is the fraction of incident photons that the CCD detects, usually expressed as a percentage. Obviously, the higher the figure the better. Consider also the QE in different filters, since most CCDs are more sensitive in the red and less sensitive in the blue.

Of equal importance for variable star work is linearity; that is, whether the recorded counts increase in proportion to the increase in incident photons. In other words, if twice as many photons hit the chip then twice as many counts should be recorded. Note that some amateur CCDs are optimised for taking pictures of galaxies, where linearity is unimportant. To check a CCD's linearity, make a series of different-length exposures of the same star and plot recorded counts against exposure time (use long exposure times, and a star that won't saturate, so that slight errors in the exposure times are minimised; this also needs to be done in cloud-free weather). If the graph deviates from a straight line at high count levels

---

Box 1.3: Image scale and field size

A CCD detector placed at a telescope's prime focus gives an image scale (the relation between the physical size of a detector and the sky coverage) determined solely by the focal length, F. A pixel with sides x microns (μm) covers θ arcsec of sky where

$$\theta = 206x/F$$

and F is in mm (the factor 206 is the number of arcsecs in a radian, divided by 1000 for the conversion from μm to mm). Thus 18 μm pixels with a 1200-mm focal length cover 3 arcsec. If the CCD is an array of 1024 × 1024 pixels the total field then covers 3072 arcsec or 51 arcmin square.

---

then the chip is non-linear, and you should use it only at lower count levels.

A similar consideration involves saturation and 'gain'. The charge storable in each pixel is limited to, say, 192 000 electrons (the 'full well'). Similarly the electronic analogue-to-digital converter (ADC) will have a limited dynamic range, and will be able to cope with up to, say, 64 000 counts. To exploit the full well capacity of the CCD, allowing longer exposures of bright stars, the ADC could be set to map three electrons onto one count (called a 'gain' of 3). However, this decreases the resolution (since it counts in batches of three photons) and so is less desirable when dealing with very low count levels. In typical observations of cataclysmics, accumulating a few hundred to a few thousand counts in a star, neither consideration is critical, but in observations of the faintest stars a gain of 1 would be best. Again, CCDs optimised for taking pretty pictures will have a higher gain.

READOUT NOISE, FRAME TRANSFER CCDS, AND DARK CURRENT

In recording the charge from each pixel the electronics introduce a 'readout noise' error of a few counts. In principle, the lower this value the better. However, lowering the value usually implies slowing down the readout and so increasing the deadtime. This is particularly important for CCDs which do not have mechanical shutters, so that light continues to fall on the CCD while it is being read out, contaminating the signal. A 'frame transfer' CCD avoids this problem, since it is divided into two halves, one of which is covered so that light cannot fall onto it. The image is recorded on one half, and then quickly shunted into the other half at the end of the exposure, from where it can be read out slowly with minimal readout noise. A further advantage (if the software allows) is that the next exposure can be accumulating while the last frame is being read out from the covered half, thus eliminating dead-time.

Since readout noise is added only once per exposure, it will be insignificant for long, well-exposed frames. It is, though, important at low count levels, for instance when a series of short exposures are made to obtain good time resolution.

A last consideration is the dark current, the excess counts due to thermal noise in the chip. Again, the lower this is the better, although practically it is only necessary for it to be lower than the sky contribution to avoid degrading the observations.

FURTHER READING ON OBSERVING TECHNIQUES

For further assistance on making observations and the equipment and software needed, consult the organisations listed in Appendix E or books such as:

Observing Variable Stars: a Guide for the Beginner, David Levy, (CUP, 1998)
Astronomical Equipment for Amateurs, Martin Mobberley (Springer–Verlag, 1999)
Astrophotography for Amateurs, Jeffrey Charles (Springer–Verlag, 2000)
The Art and Science of CCD Astronomy, David Ratledge (Springer–Verlag, 1996)
A Practical Guide to CCD Astronomy, Patrick Martinez et al. (CUP, 1997)
Star Ware, Philip S. Harrington (Wiley, 1998)

Chapter 2

The orbital cycle

Perverse as it may seem, the best place to start a discussion of cataclysmic variables is with the lightcurve of a different type of star. Cataclysmics are dominated by the flow of stellar material from one star to another, which can outshine the two stars themselves. Instead, let us consider a binary star which is very similar to a cataclysmic binary, but without the messy complication of flowing material.

The lightcurve of NN Ser is shown in Fig. 2.1. Two stars, bound together by gravity, are orbiting each other every 3 hrs 7 mins, cycling like clockwork. In contrast, Earth takes a whole year to orbit the Sun. Johannes Kepler first deduced that the orbital period of a planet is related to the radius of its orbit, which we now know to apply to all orbiting bodies. Smaller orbits have shorter periods; for instance, Mercury orbits the Sun at a third our distance, and takes only 88 days. The few-hour orbital period of NN Ser means that the two stars must be orbiting much more closely, and indeed the entire NN Ser system is smaller than the Sun! The small size means that we can never resolve such binaries, even with the *Hubble Space Telescope*, and must use indirect means to discover their nature.

Periodically, one of the components of NN Ser passes in front of the other, extinguishing the light from the system in a deep eclipse. Thus the orbital plane of the two stars must be nearly edge-on to us. We also see that the eclipsed star is by far the brighter, since when it is hidden the residual light is minimal. Further, we can deduce that the eclipsed star is tiny, because it takes only 80 secs to enter or leave eclipse. Since we know the speed at which the eclipsing star moves (from the circumference of its orbit and the orbital period) we know how far it has moved in those 80 secs, and this reveals the size of the star it has occulted. We find that the small bright star is similar in size to the Earth!

2.1 THE WHITE DWARF

A bright star the size of the Earth must be a white dwarf, the burnt-out core of a star nearing the end of its life.[1]

Normal stars are giant balls of gas, consisting mostly of hydrogen. The pull of gravity would cause the star to collapse, but is balanced by the pressure of energy

16   The orbital cycle   Ch 2

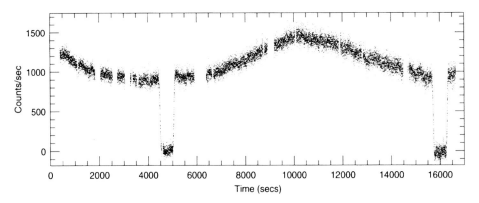

Fig. 2.1: A lightcurve of NN Ser showing eclipses every 3 hrs 7 mins. The gaps are due to cloudy weather. (Data by Janet Wood.[2])

trapped inside. In the centre, where the pressure is greatest, the gas is heated to tens of millions of degrees. Since energy is leaking away — the light we see — a star is only stable if nuclear reactions generate more. In our Sun, 500 million tonnes of hydrogen are burnt into helium every second. The ash from the nuclear furnace (helium and possibly carbon, oxygen and heavier nuclei) sinks and collects in the core, crushed to a density of a hundred-thousand times that of Earth by the weight of the surrounding layers. At such densities the electrons in adjacent atoms nearly overlap. But a result from quantum mechanics, Pauli's exclusion principle, prevents two electrons from being in the same place at the same time, forcing the atoms apart, and halting further contraction. The core of ash grows but lies inert. As hydrogen fuel becomes depleted throughout the core, the star swells into a red giant, perhaps 100 times its original radius. Pushed far from the core, the outer layers are no longer strongly bound; helped by the pressure of radiation from below they float off into space to form a planetary nebula. The hot, dense core of ash becomes exposed and is observed as a white dwarf, all that is left of the original star.

Fig. 2.2: Schematics of NN Ser's eclipse (at phases approaching eclipse, eclipse start, mid-eclipse, eclipse end, and after eclipse). The duration of eclipse ingress depends on the white dwarf size; the total eclipse duration depends on the red dwarf size. One hemisphere of the red dwarf is heated by the white dwarf's intense radiation, producing a 'reflection effect' in the lightcurve (Fig. 2.1).

Knowing that a star is a white dwarf gives us a rough estimate of its mass. No white dwarf can be heavier than the Chandrasekhar limit, about 1.4 times the mass of the Sun (1.4 $M_\odot$ in the usual notation), otherwise the pressure of gravity would force the electrons into the nuclei, where they would combine with the protons to form neutrons. The white dwarf would then collapse further into a neutron star, essentially one atomic nucleus 10 km across. The commonest mass of a white dwarf is 0.5–0.6 $M_\odot$, with almost all being found in the range 0.3–1.3 $M_\odot$.

The white dwarf in NN Ser is relatively young in astronomical terms; having thrown off its outer layers only a few million years ago it is still glowing white hot at 60 000 K, and so far outshines its companion.

2.2 THE RED DWARF

We can also measure the total duration of the eclipse, which tells us the time taken for the companion star to pass in front of the white dwarf. Knowing the speed of the star, and thus turning the time into a distance travelled, we find that the companion star must have a radius of about 0.15 that of the Sun (0.15 $R_\odot$).

Hence we know that the companion is relatively small (although still much larger than the white dwarf) and very dim. Such stars are called red dwarfs. The most common of all stars, they are smaller versions of our Sun. With less matter (a red dwarf with radius 0.15 $R_\odot$ would have a mass of 0.12 $M_\odot$) the weight of material on the core is much less and consequently the temperature in the core is lower. Nuclear reactions proceed at a feeble rate, only a thousandth the rate of those in our Sun. Consequently, the surface of the red dwarf is a cool 2900 Kelvin, compared to our Sun's 5800 K.

But note that the red dwarf, orbiting so close to the white dwarf, is blasted by the intense 60 000-K radiation, which heats the side facing the white dwarf to 7500 K. This causes the other obvious feature in the lightcurve in Fig. 2.1, the slow, smooth variation. Near eclipse, the heated hemisphere of the red dwarf faces away from us, and the cool hemisphere is too faint to be visible (hence there is no flux at mid-eclipse); as the stars orbit, more of the heated hemisphere becomes visible, until half the orbit later it is seen face-on, producing maximum brightness. Such a modulation is called a *reflection effect* (which is slightly misleading since the red dwarf absorbs and re-emits the white dwarf's radiation, rather than simply reflecting it).

Having introduced both the white dwarf and the red dwarf, I will adopt the common nomenclature of calling the white dwarf the *primary* star, and denoting it by a subscripted 1 (for example M_1 for the white dwarf mass). Similarly, the red dwarf is the *secondary*, denoted by a subscripted 2.

2.3 THE ROCHE GEOMETRY

Lone stars are spherical, pulled into the most compact configuration possible by gravity. Similarly, stars in a wide binary, where the separation is much greater than their sizes, are also spherical. In a cataclysmic variable this applies to the

Box 2.1: Keplerian motion

A modern version of Kepler's law, incorporating Newton's law of gravity, is obtained by considering a small mass m in a circular orbit around a large mass M at a radius r (by setting $m \ll M$ we can treat M as stationary). To stay in its orbit, m must experience a centripetal force (one directed towards M) of magnitude[3]

$$F = \frac{mv^2}{r}$$

which is supplied by the gravitational attraction

$$F = \frac{GMm}{r^2}.$$

Combining the two equations yields a velocity, called the *Keplerian velocity*, given by

$$v = \sqrt{\frac{GM}{r}}.$$

The circumference of the orbit is $2\pi r$ so the orbital period is $P_{\text{orb}} = 2\pi r/v$ which leads to

$$P_{\text{orb}}^2 = \frac{4\pi^2 r^3}{GM}.$$

Generalising the argument to masses of any size separated by a and orbiting their common centre of mass results in the very similar equation

$$P_{\text{orb}}^2 = \frac{4\pi^2 a^3}{G(M + m)}$$

which is referred to as *Kepler's law*.

The following three concepts will recur frequently in this book, so are given explicitly here. First, from above, the Keplerian velocity decreases as the radius of the orbit r is increased. Second, the angular momentum is given by $J = mrv$ for a mass m moving with speed v perpendicular to a lever arm r, and thus for Keplerian motion

$$J = \sqrt{GM}\, m \sqrt{r}$$

which shows that angular momentum increases as r increases. Third, the particle has gravitational potential energy P.E. $= -GMm/r$ but also has orbital kinetic energy $\frac{1}{2}mv^2$. Again using the Keplerian velocity, the total energy is thus

$$E = -\frac{GMm}{2r}$$

and so the total energy becomes more negative as r decreases.

Box 2.2: NN Ser's component stars

To find the separation a of two stars of mass M_1 and M_2 orbiting about their common center of mass we can rearrange Kepler's law (Box 2.1) to obtain

$$a^3 = \frac{G(M_1 + M_2)P_{\text{orb}}^2}{4\pi^2}.$$

To investigate the nature of NN Ser we can assume a white-dwarf mass of half the Chandrasekhar mass, knowing that it will not be wrong by more than a factor of two. Furthermore, since the companion star is so faint it is unlikely to be very massive. So let us therefore take $M_1 + M_2 = 0.7\,M_\odot$, which, with an orbital period of 3.12 hrs, gives a separation a of 6.7×10^8 m. This is slightly less than the solar radius ($R_\odot = 7 \times 10^8$ m).

As seen by the white dwarf, the red dwarf traces out a circle of circumference $2\pi a$ every 3.12-hr orbital period, and so must be travelling at a relative speed of 370 km s^{-1}. Thus the red dwarf moves 80×370 km in the 80 secs taken for the white dwarf to enter or leave eclipse, giving a white-dwarf diameter of 30 000 km. Hence the white dwarf is only 1/50th the size of the Sun and about twice the size of Earth.

The total duration of the eclipse gives the time for the red dwarf to pass in front of the white dwarf. Thus the duration of 575 secs, combined again with the relative speed of 370 km s^{-1}, gives the diameter of the red dwarf as 210 000 km (we are assuming here that we see the orbital plane exactly edge on, inclination $i = 90°$, otherwise the eclipse length would give a chord across the red dwarf and the diameter would be somewhat larger; see Fig. 2.2). Thus the red dwarf has a radius of 0.15 $R_\odot$.

Given the temperatures and radii of the components we can estimate their luminosities by assuming that they emit as *black bodies*, and thus obey the Stefan–Boltzmann law

$$L = \sigma A T^4 = \sigma 4\pi R^2 T^4$$

where L, the luminosity, is the energy per second emitted by a star of temperature T and surface area A (or radius R). σ is the Stefan–Boltzmann constant.

The white dwarf ($R_1 = 15\,000$ km, $T_1 = 60\,000$ K) thus emits 2×10^{27} W, whereas the red dwarf ($R_2 = 105\,000$ km, $T_2 = 2900$ K) emits a feeble 6×10^{23} W (for comparison our Sun emits 3.8×10^{26} W).

However, the red dwarf is a circular target of area πR_2^2 intercepting the white dwarf's radiation, which at a distance a is spread over an area $4\pi a^2$. Thus the red dwarf intercepts $\pi R_2^2 / 4\pi a^2 = 0.0062$ of the white dwarf's radiation, or 1.2×10^{25} W. This absorbed energy far exceeds the intrinsic luminosity of the red dwarf, and has to be reradiated to maintain equilibrium. By applying the Stefan–Boltzmann law to a hemisphere (area = $2\pi R_2^2$) emitting this energy we deduce that it has a temperature of 7500 K.

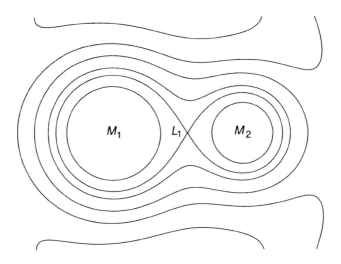

Fig. 2.3: Contours of equal gravitational potential drawn for a binary in which M_2 is half the mass of M_1. The stars orbit in the plane of the paper. The potential surfaces just touching at the L_1 point are the Roche lobes of the two stars.

white dwarf (which has a radius of $1/50^{\text{th}}$ the binary separation) but not to the much larger red dwarf. Instead, the lighter, less dense secondary is distorted by the gravity of its close companion, which pulls at the fluffy outer layers. If one imagines moving the two stars closer, the secondary becomes increasingly distorted until the material nearest the primary experiences a greater gravitational attraction towards that star than back towards its own star. Material then flows between the two stars, the characteristic of a cataclysmic variable.

Because of its importance to binary systems, the outline of a star at the critical point when it is just possible for mass to transfer to the companion is given a name, the *Roche lobe*. The apex of the Roche lobe, called the *inner Lagrangian point* (abbreviated to L_1), is the easiest path by which material can transfer between the stars. Imagine the gravitational field of the two stars as deep valleys, into which water tends to flow. The L_1 point is the lowest mountain pass between the valleys.

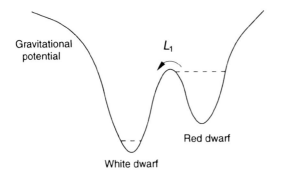

Fig. 2.4: The Roche lobes can be visualised as gravitational wells in which matter flows downhill. When the red dwarf overfills its Roche lobe, matter flows over the L_1 point into the Roche lobe of the white dwarf.

If one valley were dammed, water would fill up the valley until it lapped over the pass, whereupon any further water would flow into the second valley. Similarly, if the amount of material in a star is too much for the Roche lobe, it overflows onto the companion star.

In NN Ser the red dwarf is slightly smaller than the Roche lobe, so nothing happens; but in cataclysmic variables the mass transfer due to a red dwarf filling its Roche lobe gives rise to the some of the most diverse and heavily studied phenomena in astrophysics.

One immediate consequence is that a red dwarf that fills its Roche lobe spins at the same rate as it orbits, called *tidal locking*. If this were not the case, there would be continual tidal flows of material into and out of the bulges of the Roche lobe, expending enormous energy. Thus the star quickly adjusts its spin period to match the orbit, so that the same material remains in the bulges. For the same reason the Moon is tidally locked and always presents the same face to Earth.

2.3.1 Ellipsoidal variations

The distorted Roche-lobe shape of the red dwarf is often detectable directly in the lightcurves of cataclysmics. By looking in infrared light we maximise our sensitivity to the cool secondary while minimising the light from the hotter components. Fig. 2.6 shows an infrared lightcurve of XY Ari. The distorted secondary presents maximum surface area to us (and hence appears brightest) when it is seen side on, which occurs a quarter of a cycle before or after eclipse. But when its elongation lies along our line of sight it is fainter. This is called an *ellipsoidal modulation*.

Modelling the ellipsoidal modulation — using our knowledge of the Roche geometry to predict the ellipsoidal modulation from a particular binary and then comparing this to the observed lightcurve — is a way of deducing the parameters of a cataclysmic variable. It is sensitive, firstly, to the relative masses of the two stars, since the smaller the mass of the secondary compared to the primary the more distorted it is. In fact the *shape* of the Roche lobes is determined solely by the mass ratio, q, defined by $q = M_2/M_1$. The *scale* of the Roche lobe is then set by the size of the orbit. Secondly, the ellipsoidal variation depends on the inclination of the binary orbit: if the orbit is seen edge on (as in XY Ari) the variation in observed surface area is large and so the ellipsoidal modulation is large; in a lower inclination binary the effect is reduced, until when seeing the stars orbiting face on (as in the view of Fig. 2.3) the Roche lobe presents equal area at all times and there is no ellipsoidal modulation.

2.4 MASS TRANSFER

From a small patch at the Lagrangian point, covering about a thousandth of the red dwarf's surface, emerges a thin stream of stellar material. Pushed from behind by the pressure of the stellar atmosphere, it finds itself in the empty space of the primary's Roche lobe. Naively, one expects it to fall towards and onto the primary, but on consideration one realises that it cannot do this directly. The material

Box 2.3: The Roche geometry

An entirely general analysis of the geometry of orbiting stars is complex enough to require a computer model. However, the problem can be simplified (along the lines pioneered by Edouard Roche in the nineteenth century) by assuming that tidal forces have coerced the stars into a circular orbit and that the mass of each star can be treated as concentrated at the star's centre (both of which are good approximations). From here, consider the set of hypothetical surfaces over which the gravitational potential is constant ('equipotentials'). Near one of the stellar centres the equipotentials are circular, but on scales comparable to the separation the tidal force distorts them into 'pear' shapes pointing towards the companion star (see Fig. 2.3). The centrifugal force due to the orbital motion also flattens the equipotentials into the orbital plane.

We can write the gravitational potential Φ at a point specified by the vector r as the sum of the potentials of the two stars (with masses M_1 and M_2 located at r_1 and r_2 respectively) and a third term due to the centrifugal force. Thus

$$\Phi = -\frac{GM_1}{|r - r_1|} - \frac{GM_2}{|r - r_2|} - \frac{1}{2}(\omega \wedge r)^2$$

where ω is the angular frequency of the orbit.

The equipotentials for several values of Φ are plotted in Fig. 2.3. As Φ is increased, the equipotentials of the two stars first touch and then merge. The critical surface when the equipotentials first touch is called the *Roche lobe*. The point at which they touch is the inner Lagrangian point, or L_1 (there are other Lagrangian points of less significance). If neither star fills its Roche lobe (as in NN Ser), the binary is described as *detached*; if one star fills its Roche lobe the binary is described as *semi-detached* (all cataclysmics are semi-detached); if both stars fill or overfill their Roche lobes the binary is said to be in *contact* (W UMa stars are examples of contact binaries).

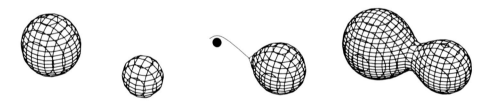

Fig. 2.5: *Left:* A detached binary with both stars within their Roche lobes. *Middle:* A semi-detached binary: the secondary fills its Roche lobe emitting a stream of material from L_1. If the primary is small enough, the stream will orbit around it. If it were larger, the stream would hit the primary, as occurs in some Algol-type binaries. *Right:* A contact binary, with both stars overfilling their Roche lobes.

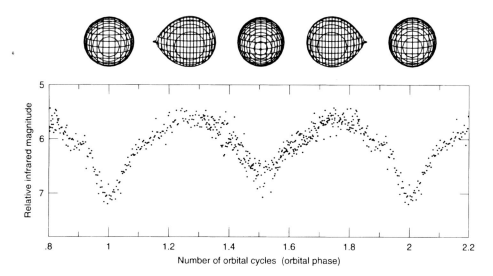

Fig. 2.6: The infrared lightcurve of XY Ari showing ellipsoidal variations. The Roche-lobe-shaped secondary is seen side-on twice per orbital cycle, presenting maximum surface area and thus appearing brighter. When it is end-on (L_1 pointing towards or away from us) XY Ari is dimmer. Schematics of the secondary at the different phases are shown above the figure. Note that XY Ari has a cooler white dwarf than NN Ser so there is no reflection effect. The figure uses the standard convention of plotting against orbital cycle, where phase increases by 1 each cycle, and the white dwarf is furthest from us at phases 0, 1, 2, and so on.[4]

squirts out of the L_1 point at roughly the sound speed in the gas (this is because the particles escape because of their thermal motion, and it is the same thermal motion that carries sound). This speed is ~ 10 km s^{-1}. But the L_1 point itself is orbiting perpendicular to this motion, at more than 100 km s^{-1}. Injected with this orbital motion, the stream swings into an orbit around the white dwarf rather than flowing directly towards it (an equivalent statement, considering the motion from a frame rotating with the binary, is that the Coriolis force causes the stream to swirl around the white dwarf, just as the Earth's rotation produces the swirl of a hurricane).

Following a trajectory determined by the injection velocity and the gravity of the primary (the gravity of the secondary quickly becomes insignificant), the stream sweeps past a close approach to the white dwarf, and loops round to cross its earlier path (see Fig. 2.7).[5]

What happens next? A full account of the stream hitting itself involves messy, turbulent flow that can be addressed only by a computer simulation. But we can deduce the result by realising that the turbulent shocks dissipate energy, and so the material will settle into the lowest-energy type of orbit: a circular one. However, the stream cannot so easily reduce its angular momentum, and thus the material orbits at a radius that ensures that it has the same angular momentum as material

> **Box 2.4: Formulae for the Roche geometry**

The Roche geometry is completely specified by the mass ratio q ($\equiv M_2/M_1$) and the separation a (obtained from Kepler's law, Box 2.1). However, it does not lead to simple formulae for items such as the distance to L_1, etc. Instead, one has to calculate the full Roche geometry on a computer, and distill the results into formulae that are approximate but sufficiently accurate. The following are commonly used and are accurate to $\approx 1\%$ (see Warner 1995 for a fuller account).

The distance of L_1 from the primary is given by

$$R_{L_1} = a\,(0.500 - 0.227 \log q) \qquad \text{for } 0.1 < q < 10.$$

The radius of the secondary varies, owing to its distorted shape, but a sphere containing the same volume as the Roche lobe has a radius

$$R_2 = a\,0.462 \left(\frac{q}{1+q}\right)^{1/3} \qquad \text{for } 0.1 < q < 0.8$$

or alternatively

$$R_2 = \frac{a\,0.49 q^{2/3}}{0.6 q^{2/3} + \ln(1 + q^{1/3})} \qquad \text{for all } q.$$

The distance of closest approach of a free-fall stream to the primary is given by

$$r_{\min} = a\,0.0488\,q^{-0.464} \qquad \text{for } 0.05 < q < 1$$

and the circularisation radius is

$$r_{\text{circ}} = a\,0.0859\,q^{-0.426} \qquad \text{for } 0.05 < q < 1$$

which is also the minimum radius of the disc's outer edge. The maximum disc radius before it is truncated by the tidal action of the secondary is

$$r_{\text{tidal}} = a\,0.60/(1+q) \qquad \text{for } 0.03 < q < 1.$$

A less accurate but physically instructive estimate of the circularisation radius is obtained by arguing that the angular momentum of the stream is conserved. The specific angular momentum of material at the L_1 point is $R_{L_1} v$ where the velocity v at R_{L_1} is $2\pi R_{L_1}/P_{\text{orb}}$. The specific angular momentum after circularisation is $r_{\text{circ}} v_{\text{Kep}}$ where the Keplerian velocity v_{Kep} is $\sqrt{GM_1/r_{\text{circ}}}$. Equating angular momentum then gives $r_{\text{circ}} = 4\pi^2 R_{L_1}^4 / GM_1 P_{\text{orb}}^2$, and eliminating P_{orb} by using Kepler's law (Box 2.1) leads to

$$r_{\text{circ}} = (1+q)\,R_{L_1}^4/a^3.$$

Sec 2.4   Mass transfer   25

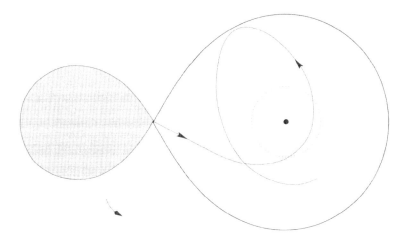

Fig. 2.7: A plan view of the trajectory of a gas stream emanating from the secondary (left). The primary nestles deep within its Roche lobe (outer line). The stream collides with itself and ultimately forms a ring at the circularisation radius (dotted line). Note that the trajectory is as seen by someone orbiting with the binary (the 'rotating frame'), thus one should envisage the whole diagram rotating while the stream traces the trajectory. The figure is drawn for a q of 0.3.

at L_1. This is known as the *circularisation radius*.

To understand what follows bear in mind three concepts. First, material in a smaller orbit moves faster (from Kepler's law). Second, material in a smaller orbit has a lower angular momentum (the increase in speed is not enough to offset the decrease in radius). Third, by transferring into a smaller orbit, material liberates gravitational energy.

Thus, within the ring of material orbiting at the circularisation radius, blobs of material slightly nearer the primary will orbit slightly faster, causing friction as they slide past blobs further out. The friction and turbulence heat the gas so that energy is radiated away. The energy comes from the gravitational field, which means that some of the material must have moved to smaller orbits in the process. But to conserve angular momentum, other material must move to larger orbits. Thus, overall, the ring spreads out into a thin disc. The disc continues to spread until the inner edge meets the primary. Material continually flows through the disc, spiralling inwards on ever smaller orbits, and so accretes onto the white dwarf.

Angular momentum flows outwards through the disc, enabling the inward flow of material and the consequent release of energy. At the outer edge of the disc, tidal interactions with the secondary star soak up the angular momentum and return it to the orbit of the secondary. This limits the outward spread of the disc.

The disc is replenished by the mass-transfer stream from the secondary, which brings both fresh material and more angular momentum. As long as it flows, the disc is maintained.

2.4.1 The accretion disc

The thin disc of circling material, destined to settle onto the compact star lurking at its centre, is called an *accretion disc*. The story of cataclysmic variables is primarily the story of accretion discs, though this is deferred for now to take centre stage in Chapters 5 and 6. It is also the accretion disc that accounts for much of the interest of professional astronomers in these stars, since discs are not limited to cataclysmics but are widely spread.

Around newly forming stars an accretion disc performs the same task as in a cataclysmic binary, soaking up angular momentum to allow a gas cloud to collapse into a star. In *Hubble Space Telescope* images of star-forming regions, accretion discs are seen as cool, dark bands hiding young stars. And it is from such discs that planets form. When the Solar System was young, Earth coalesced from such a disc, whose remnants can still be found in the belt of asteroids orbiting between Mars and Jupiter, and in the meteors seen when specks of dust burn up in Earth's atmosphere. One look at the cratered surface of the Moon shows how much more was orbiting when the planets first formed.

At the other end of the Universe, it is accretion discs that power the quasars.

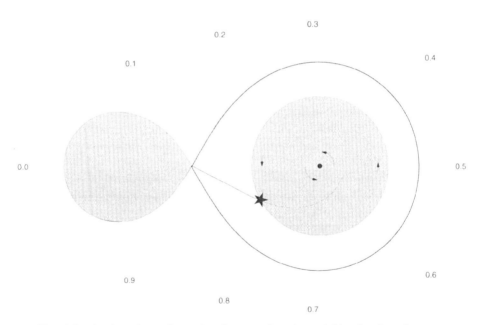

Fig. 2.8: A plan-view schematic of a cataclysmic variable showing the mass-transfer stream colliding with the disc at a 'bright spot'. Sometimes, part of the stream overflows the disc and continues to near the white dwarf (dotted line). The arrows on the disc illustrate the fact that material orbits faster near the disc centre. The tickmarks surrounding the figure indicate the direction from which the binary is viewed at the orbital phases labelled, assuming it to be seen edge-on. For lower inclinations, imagine the viewpoints lifted out of the plane of the paper.

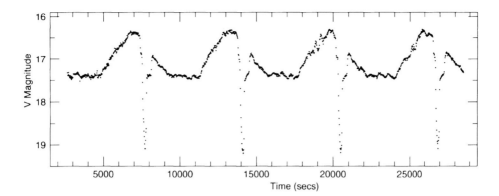

Fig. 2.9: A lightcurve of IY UMa showing prominent 'orbital humps' along with deep eclipses. (Data by the Center for Backyard Astrophysics.[6])

In the centre of distant galaxies, monster black holes, weighing perhaps 100 million solar masses, are surrounded by accretion discs so bright that they outshine the host galaxy. They are the farthest things we can see. The discs are fed by whole stars which stray too close and are shredded by the powerful tides generated by the colossal gravitational fields, to become fodder for the black hole.

But quasars are at vast distances, and young stars are shrouded by the dust from which they formed, so nowhere can we see and study accretion discs as clearly as we can in cataclysmic variables. Thus the knowledge gained from monitoring the lightcurves of cataclysmics — discovering the workings of accretion — can be applied to the formation of new planets and to the powerhouses of the Universe's most energetic phenomena.

2.4.2 The bright spot

Once a disc has formed, the stream from the secondary hits the disc edge, forming a *bright spot*. Here, the stream material, falling radially, encounters disc material moving across its path on a circular orbit. The details of the turbulent encounter are poorly understood, but computer simulations suggest that the dense core of the stream can punch a hole in the disc, being only gradually assimilated into the circular flow; and also that the stream is slightly wider than the disc edge, so that some stream material flows over the disc, continuing on its original trajectory.

The kinetic energy of the stream is converted to heat and radiated away during the encounter, so that in some cataclysmics the collision region emits $\sim 30\%$ of the total light of the system. We know this through the observation of *orbital humps* (Figs. 2.9 and 2.11). They are caused by the additional light seen when the bright spot is on the side of the disc facing us, peaking around phases 0.8–0.9 (see Fig. 2.8).

As we saw with NN Ser, eclipsing systems offer the best opportunity to understand cataclysmics. The time at which a feature is eclipsed ties down its location in the binary, and the amount of light eclipsed tells us how bright it is (see Fig. 2.10)

28 The orbital cycle

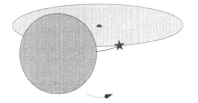

Fig. 2.10: The eclipse in a high-inclination cataclysmic binary. The secondary is partially eclipsing the disc, is about to eclipse the white dwarf, and will then eclipse the bright spot.

To exploit this opportunity fully, many eclipse observations can be averaged to produce a very high quality lightcurve (Fig. 2.11). From such data it is apparent that the eclipse contains several parts. The steepest sections are caused by the white dwarf entering eclipse (phase 0.97) and emerging again (phase 1.03); as in NN Ser, the smallness of the white dwarf means that these happen very rapidly. We also see the bright spot becoming eclipsed (phase 0.99) and emerging again (phase 1.08). These take slightly longer, telling us that the bright spot is somewhat larger than the white dwarf. In addition, there is a slower eclipse of the accretion disc, lasting from phase 0.92 to 1.08, which tells us (no surprise) that the disc is much bigger than either the white dwarf or the bright spot.

One can use such deductions to go further and separate the lightcurve into components from the white dwarf, the bright spot and the disc (Fig. 2.12). Indeed, such techniques reveal much of our information about cataclysmics. The phase at which the bright spot is eclipsed, given that it must lie along the known trajectory of the stream, tells us the radius of the disc. Using this, one finds that the smallest discs fill about half the primary's Roche lobe. The largest discs fill $\sim 80\text{--}90\%$ of the primary's Roche lobe, where they are truncated by tidal interactions with the secondary star.[7]

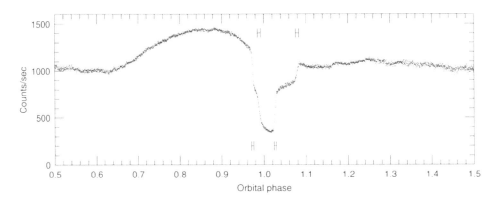

Fig. 2.11: Data from 17 eclipses of Z Cha combined into one lightcurve by 'folding' on orbital phase. The H marks below the data locate the episodes when the white dwarf enters and then leaves eclipse, and those above the figure are the same for the bright spot. The orbital hump is clearly seen between phases 0.65 and 1.1, though it is eclipsed for some of this time. (Lightcurve by Janet Wood.[8])

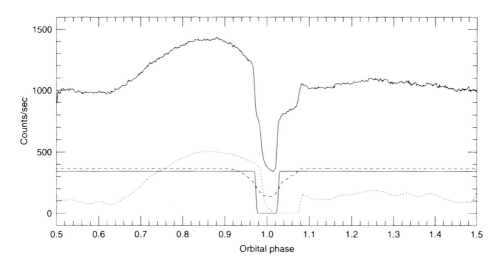

Fig. 2.12: The lightcurve of Z Cha (top) decomposed into the lightcurve of the white dwarf (solid line), reminiscent of that in NN Ser, and the lightcurve of the bright spot (dotted), including the orbital hump and its eclipse, and the lightcurve of the disc (dashed) showing its broader eclipse. This is shallower, since the red dwarf is not big enough to completely hide the disc. (Analysis by Janet Wood.[8])

Furthermore, the duration of the white-dwarf eclipse depends only on the Roche-lobe geometry, and thus provides information on the mass ratio, q, and the inclination of the system, i. From here, only a few pieces of the jigsaw are needed for a complete knowledge of the binary parameters, and these are discussed in the next chapter.

2.4.3 Grazing eclipses

We complete this chapter with a variation on the eclipse theme. Fig. 2.13 shows the lightcurve of U Gem, complete with orbital hump and an eclipse. However, the eclipse has only one component, unlike that of Z Cha, and is not sharp-edged enough to be that of a white dwarf. The interpretation is that U Gem has a slightly lower inclination than Z Cha (70° instead of 82°)* so that the eclipse is grazing (imagine moving the secondary down a little in Fig. 2.10). We therefore see an eclipse of the bright spot but not of the white dwarf.

A similar eclipse of the bright spot only is seen in WZ Sge (Fig. 2.14). However it also shows a peculiar double-peaked hump which is bright at phases 0.2–0.4 as well as the usual 0.7–0.9. There is no settled interpretation of this lightcurve. One possibility is that it is an ellipsoidal modulation, similar to that in XY Ari (Fig. 2.6). However, the double-hump in WZ Sge is seen in blue light, whereas the secondaries (and thus the ellipsoidal modulations) are bright only in the infrared (see the next

---

*The inclination, i, is specified by an angle between 0° (face-on) and 90° (edge-on).

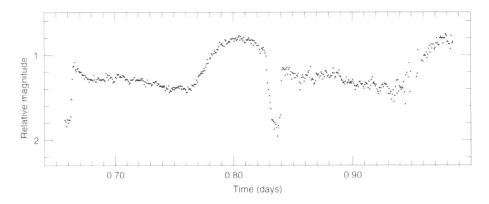

Fig. 2.13: U Gem's lightcurve showing an orbital hump and an eclipse of the bright spot. (Data by the Center for Backyard Astrophysics.)

chapter for the reason). Furthermore, the amount of 'flickering' in the lightcurve is not characteristic of the smooth geometrical effect of an ellipsoidal modulation.

The most probable explanation is that the humps are again caused by the bright spot, but that in this star the accretion disc is transparent, so that the bright spot can be seen through the disc, producing a second hump when it is furthest from us. In addition, the dip at phase ~ 0.3, cut into the anomalous hump, could result if the central regions of the disc are only partially transparent, so that we see less light when the bright spot is directly behind the centre of the disc.

The transparency of the disc in WZ Sge could result from a very low rate of mass transfer from the secondary star in this system.

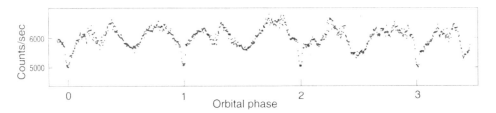

Fig. 2.14: WZ Sge's lightcurve, showing an eclipse of the bright spot and a peculiar double orbital hump. The data, obtained by Joe Patterson, are 20-sec integrations in blue light with a 2.1-m telescope.[9]

Chapter 3

Spectral characteristics

Objects emit light when they are hot. Thus fires, the Sun, light-bulb filaments, and so on, produce the light we live by. There are other ways of creating electromagnetic photons (we exploit this to generate radio waves for communication) but most photons in the Universe result from thermal energy.

The characteristics of this radiation are set by fundamental physics, and are summed up in an idealised entity called a *blackbody*. Such a body absorbs all the light that falls on it and emits the maximum radiation allowed for its temperature. Although no natural thing is a genuine blackbody,[*] most are near enough that blackbody behaviour is a good approximation.

Max Planck first derived the distribution with wavelength of the radiation emitted by a blackbody and found that it depends only on its temperature (see Fig. 3.1). Furthermore, the total energy emitted across all wavelengths rises dramatically with temperature, scaling as the fourth power. As temperature increases, the bulk of the extra radiation emerges at shorter wavelengths (more energetic photons) and so the peak of the distribution shifts shortward.

At the temperature of our Sun, 5800 K, the radiation peaks at wavelengths to which our eyes are sensitive — they have evolved to exploit the most abundant supply of photons. The light from a lower-temperature red dwarf ($T \sim 3000$ K) peaks in the infrared and falls off rapidly at bluer wavelengths, emitting essentially nothing in the ultraviolet. The much hotter white dwarf ($T > 12\,000$ K) pumps out most of its energy in the ultraviolet with less in the red; a still hotter body would radiate mostly X-rays.

3.1 THE WHITE DWARF SPECTRUM

A white dwarf is one of the nearest approximations to a blackbody (see Fig. 3.2). Being hot, its flux rises steeply at bluer wavelengths. The departures from the blackbody curve are caused by a thin layer of hydrogen atoms (or sometimes helium) on its surface. In accordance with quantum mechanics, electrons move around the nuclei of the atoms in set orbits with set energies. Thus a photon emerging from the

---

[*]With the possible exception of black holes.

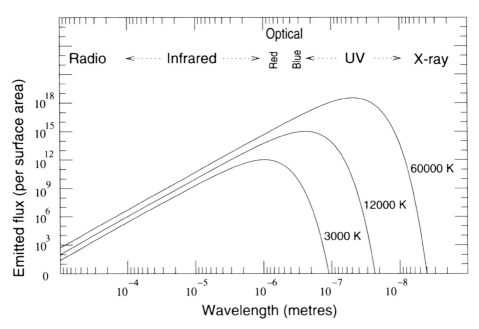

Fig. 3.1: Blackbody radiation curves at three temperatures. Note how small a portion of the electromagnetic spectrum can be seen by our eyes. The y-axis scale is logarithmic, with each large tick denoting a thousand-fold increase; thus radiation increases hugely with temperature (total flux $\propto T^4$). As temperature increases, the peak shifts to shorter wavelengths (which leads to Wien's law for the peak wavelength, λ_{peak}, given by $\lambda_{\text{peak}} T = 2.9 \times 10^{-3}$ m K). Note, though, that a hotter body emits more light per surface area at *all* wavelengths.

hot core of the white dwarf might have the right energy to bounce an electron from one orbit to another. The photon is absorbed, giving up its energy to the electron, and the emergent spectrum has a deficit at the wavelength corresponding to that energy. This is seen as a series of notches, known as *spectral lines*. The rest of the spectrum is referred to as the *continuum*. The lines in the optical region of the spectrum in Fig. 3.2 are caused by electrons that originated in the second-lowest orbit of a hydrogen atom, and are called 'Balmer' lines; at far left in the UV are more energetic 'Lyman' lines caused by electrons that originated in the lowest orbit.[1]

The white-dwarf spectrum is also notable for the width of the spectral lines, which are far broader than the corresponding features in normal stars. This is due to a phenomenon called 'pressure broadening'. Because of the intense gravity on the white dwarf surface (100 000 times that on Earth) the pressure in its atmosphere is vast; atoms are constantly being jiggled by their neighbours so that they can exist in one energy state for only a short time before being disturbed. In accordance with Heisenberg's uncertainty principle, this produces uncertainties in the energy of the electron orbits. Hence photons with a larger range of wavelengths may find an atom to absorb them, so that the absorption lines are broadened out.

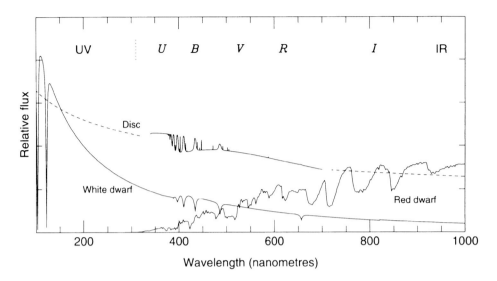

Fig. 3.2: Schematic spectra of the components of a cataclysmic variable. The hot white dwarf rises strongly at shorter wavelengths while the cooler red dwarf is brightest in the infrared. The dashed line is an approximation to a disc spectrum, created by summing up the radiation from blackbodies of different temperatures. In the optical region the dashed line is replaced by the spectrum from a disc model, complete with emission lines (computed by Martin Still). The spectrum we observe is, of course, the sum of the three components. This is normally dominated by the disc, which is both big and hot, although the relative proportions depend on the particular system. The labels $UBVR$ and I show the regions of spectrum recorded by the different filters. Only the B, V and R light is visible to our eyes; regions to the left of the vertical dotted line are absorbed by the atmosphere and are unobservable except with satellites.

3.2 THE RED DWARF SPECTRUM

As befits a much cooler body, the red dwarf's spectrum is significant only at red and infrared wavelengths. In fact, from Fig. 3.1 one might suppose that the cooler body should be lost in the white dwarf's glare at all wavelengths. Its saving grace, though, is that blackbody emission increases in proportion to the surface area, and the area of the red dwarf is a thousand times that of the white dwarf, enabling it to be seen in red and infrared light.

The red dwarf's spectrum also deviates from the blackbody form. This time, the dominant features are caused by molecules rather than atoms. The low-energy bonds binding atoms into molecules can survive in the low-temperature atmosphere of a red dwarf but not in a hotter white dwarf. These molecules again absorb photons emerging from the core. However, since molecules can rotate and vibrate in many more ways than can electrons in atoms, the features in the resulting spectrum are a complex set of broad dips. By a quirk of nature, most of the absorption dips

34 Spectral characteristics

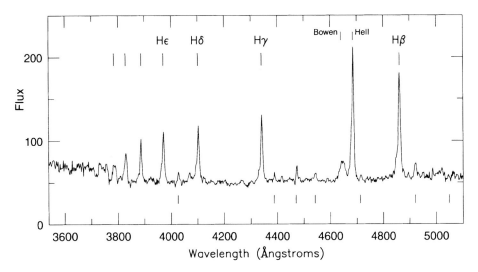

Fig. 3.3: This spectrum of TV Col shows mostly disc emission.[2] The most prominent series of lines is the 'Balmer' series of hydrogen, which are labelled Hα, Hβ, Hγ, and so on (Hα is at 6562Å and so not on this figure). There is also a strong line due to ionised helium (He II) at 4686Å and a 'Bowen fluorescence' feature at 4640Å, caused by ionised carbon and nitrogen. There are weaker lines due to neutral helium (He I) marked by ticks at 4026, 4388, 4471, 4713, 4921 and 5048Å, while the line at 4542Å is He II.

in the red-dwarf spectrum are caused by the extremely rare titanium oxide molecule (TiO), which just happens to be exceedingly efficient at absorbing photons.

3.3 THE ACCRETION DISC SPECTRUM

While the spectra of stars are well studied and well understood, accretion discs are more problematic. In part this is because discs in cataclysmic variables are relatively cool at their outer edge ($\sim$ 5000 K), but heat up through the release of gravitational energy to $\sim$ 30 000 K in their inner regions, so they cannot be treated as a single entity. The simplest approach to disc spectra is to assume that each annulus in the disc emits as a blackbody of the appropriate temperature, and to add up blackbody spectra for each annulus (weighting by its area). The range of temperatures produces an overall spectrum that is flatter in the optical than those of the red and white dwarfs (see Fig. 3.2). While the blackbody approach gives a rough approximation to the continuum emission of discs, it cannot reproduce the line features seen in their spectra (e.g. Fig. 3.3).

An alternative approach is to assume that each annulus in the disc emits the same spectrum as a star of that temperature, and so synthesise a disc spectrum by adding up a series of stellar spectra with the relevant range of temperature. Again this works approximately and includes line features. However, the lines observed from discs are often in *emission*, whereas stellar spectra show absorption features

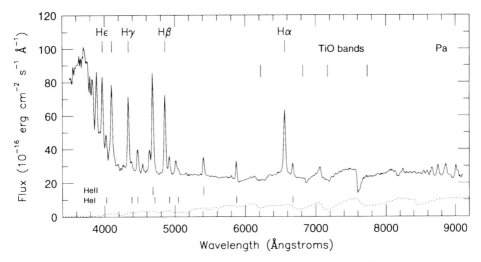

Fig. 3.4: A spectrum of RX 1313−32 showing TiO absorption bands from the secondary star, coupled with a hot emission-line spectrum from the accretion flow. In most cataclysmics the spectral features of the secondary star cannot be seen in optical spectra because the disc light dominates; RX 1313−32, though, has no disc (see Chapter 8). The main emission lines visible are the Balmer series; another series of hydrogen lines called the Paschen series, at > 8600Å; and lines due to He I and He II. The feature at 7600Å is caused by the atmosphere. The dotted line is the red-star spectrum reproduced from Fig. 3.2 for comparison. (Data courtesy of Hans-Christoph Thomas.[3])

only.[†] To understand the difference we need to discuss the formation of spectral lines in some detail.

3.3.1 Emission lines versus absorption lines

If an atom absorbs a photon by jumping up an energy level it is very likely to then jump back down and re-emit an exact replica of the first photon. In an equilibrium state ('local thermodynamic equilibrium' or LTE) the number of absorptions equals the number of emissions. This will occur when material is *optically thick*; that is, when absorption is so likely that photons can travel only a short distance between interactions with atoms (their 'mean free path' is short) and can escape the emitting material only after many such interactions. In optically thick material at a uniform temperature the absorption cancels the emission and no spectral lines are seen — the emitted spectrum is that of a blackbody.

Now consider photons entering an optically thick gas cloud. Before they are absorbed they will penetrate (on average) to a distance equal to the mean free path (this location is called an 'optical depth' of one). Similarly, photons emerging from the cloud came from an optical depth of one. But because photons at the wavelength

---

[†]Well, almost always.

of a spectral line are ready-made to interact with atoms (whereas photons at other wavelengths are absorbed only by less efficient processes) their mean free path is shorter. Thus the line photons we see emerge from nearer the surface and continuum photons emerge from deeper.

This would still make no difference in an isothermal emitter (the spectrum would still be a blackbody); however, stars and discs are hotter inside and cooler towards their surfaces. Since the continuum comes from deeper, where it is hotter, the radiation will be brighter. The radiation in the lines, coming from the cooler surface, is dimmer. Thus the lines appear as absorption features.

The opposite case of very little absorption is called *optically thin* conditions, which means that all emitted photons emerge from the emitting cloud unscathed. Two processes can still boost electrons into energetic orbits. They are, firstly, collisions of atoms ('collisional excitation'), and secondly, energetic radiation shining onto the cloud and knocking electrons out of atoms ('photo-ionisation'). When the electrons decay back to the lowest-energy orbits they produce line emission. In the absence of absorption, we see both the line emission and any continuum radiation, and hence record an emission-line spectrum.

The point of the preceeding discussion is that accretion discs are observed to show absorption lines on some occasions and emission lines on others. Thus they must be changing state between optically thick and optically thin conditions. Sometimes they even show absorption and emission features simultaneously! This can occur if an optically thick disc (absorption spectrum) is surrounded by a hot, optically thin 'corona' (emission spectrum).

Given all the complications, there is no easy way of modelling an accretion-disc spectrum. A full description needs to take into account the pressure and density

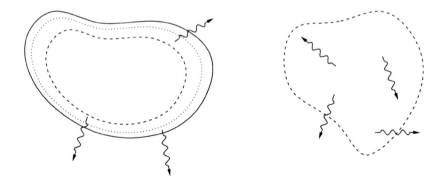

Fig. 3.5: In an optically thick gas cloud (*left*) the photons emerge from near the surface. Since the line photons interact more strongly with atoms, their 'mean free path' is shorter and (on average) they arise from nearer the surface (dotted line) than do the continuum photons (dashed line). If the cloud becomes cooler towards its edges, the line photons will reflect cooler conditions and so be fainter against a brighter continuum; hence we see absorption lines. In optically thin conditions (*right*) line photons arise from anywhere and are not absorbed. We see the line emission in addition to the continuum, and hence see emission lines.

Sec 3.4   Doppler shifts 37

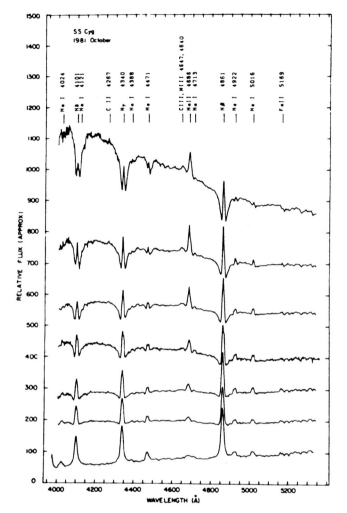

Fig. 3.6: A compilation of spectra of SS Cyg while it was fading, illustrating the different spectra in different states. As it fades, SS Cyg's disc becomes cooler (note the flatter spectrum in the lower data) and absorption features in the lines disappear. The main line features are labelled with the ion species responsible and the wavelength. (Figure by Rick Hessman.[4])

in different parts of the disc (which depends of the gravity of the white dwarf), the generation of radiation by viscous processes, the propagation of the radiation through the disc material, and additional complications such as turbulence in the gas and the transport of energy by convection. The model disc spectrum in Fig. 3.2 attempts to include all these aspects, but such work is 'cutting edge' and resemblances to observed spectra are still only approximate.[5]

3.4 DOPPLER SHIFTS

When we look at spectral lines from an accretion disc we see a shape that is created by the Doppler effect. This shifts the wavelength of photons being emitted by moving material. At root, the wavelengths of line photons have set values, since they

Box 3.1: Accretion disc temperatures

In obtaining a rough estimate of the temperatures of accretion discs it can be assumed that the gravitational potential energy released by material spiralling inward is radiated away as blackbody radiation. Given that the potential energy, P.E., of a mass m at a distance r from a mass M is given by $-GMm/r$, the energy released as m spirals in the short distance from radius r to $r-dr$ is

$$\Delta \text{P.E.} = -GMm\left(\frac{1}{r} - \frac{1}{r-dr}\right).$$

We can convert this into the *rate* of energy release by replacing m with the rate of flow of matter, dm/dt, abbreviated to $\dot{m}$.

The annulus of thickness dr has an area of $4\pi r\, dr$ (note that it has two sides) and so (using the blackbody assumption; see Box 2.2) has an energy output of

$$4\pi r\, dr\, \sigma T^4.$$

Equating the two energy equations [and multiplying through by $r(r-dr)$ and simplifying] leads to

$$GM\dot{m} = 4\pi r^2(r-dr)\sigma T^4$$

whence setting $r-dr \approx r$ gives

$$T^4 = \frac{GM\dot{m}}{4\pi r^3 \sigma} \quad \text{or} \quad T \propto r^{-3/4}.$$

In deriving this equation I have ignored two effects. First, the Keplerian orbital velocity increases at smaller radii so that roughly half the gravitational energy goes into the kinetic energy of the disc material as it spirals inward; thus the right-hand side of the above equation should be divided by two. The second effect arises because the accretion disc joins onto the white dwarf, which is generally rotating far more slowly than the inner disc. This transition region is called the *boundary layer*, and extends to ~ 1 white-dwarf radius above the surface. Here the material is slowed to the speed of the white dwarf, and the kinetic energy is turned into thermal energy and radiated away; as much as half the total accretion luminosity emerges from the boundary layer. Some of this energy is convected into the outer disc, raising its temperature. When both effects are taken into account, the factor of $1/4\pi$ in the last equation becomes $3/8\pi$, but the rest, including the $T \propto r^{-3/4}$ law, remains valid.[6] As a consequence of convective energy loss, the temperature in the boundary layer is lower than given by the above equation (the details depend on the temperature and rotation rate of the white dwarf).

In order to put numbers into the equation we can use 0.7 $M_\odot$ as a typical white-dwarf mass and 10^{14} kg s^{-1} as a typical rate of mass flow through a disc (see Chapter 4). A disc radius near the tidal limit, say 3×10^8 m, then gives an outer disc temperature of ≈ 5000 K. One white-dwarf radius above the white-dwarf surface ($r \approx 1.5\times 10^7$ m), the inner-disc temperature is $\approx 50\,000$ K.

Box 3.2: Eclipse mapping

Do observations of discs confirm the $T \propto r^{-3/4}$ law predicted by theory (Box 3.1)? We can investigate by using a technique developed by Keith Horne called *eclipse mapping*.[7] Fig. 2.10 shows that at each stage of an eclipse some parts of the disc are disappearing behind the secondary star limb and some are reappearing. Thus, if we knew the temperature of each part of the disc, we could calculate the spectrum of light emitted by the regions about to appear or disappear (assuming blackbody radiation) and so predict the changes over a small phase interval. Thus we could predict the eclipse lightcurves in several different colours, say $UBVR$ and I.

The eclipse mapping technique starts by guessing the temperature of the different parts of the disc, predicting the consequent eclipse lightcurves in several bands, and then comparing the predictions with actual data. It then alters the disc temperatures until a good match between the predicted and observed lightcurves is obtained.

The main difficulty is that the eclipse yields one-dimensional information — disc regions all along the secondary limb, with a range of temperature and radii, combine to give only one bit of information, the change in observed light — whereas the disc is two dimensional. Thus there are an infinite number of temperature distributions across a disc that would result in the same eclipse profile. To obtain a result, one chooses the smoothest of the distributions that match the data ('maximum entropy'), hoping that nature has not made the disc more patchy than necessary. A lesser difficulty is that to derive accurate temperatures one needs to know the distance to the star, so that one can calculate how bright each segment of the disc would be as seen from Earth. Distances to any astronomical objects are often poorly known (see Box 4.2), but fortunately a reasonable error in the distance has a relatively minor effect on an eclipse map.

Fig. 3.7 shows the temperature of the disc in Z Cha in two different observations. In one the data fit the $r^{-3/4}$ law very well. The other, though, shows a temperature that is nearly constant with radius, in violation of the theory. The solution is that the disc was not then in a steady state, but that material was building up in part of the disc, rather than flowing smoothly through it. This is discussed further in Chapter 5.

Further difficulties in applying eclipse mapping include: (1) To analyse the eclipse of the disc one first needs to remove variations due to the orbital hump or the eclipse of the white dwarf, performing a deconstruction similar to that in Fig. 2.12. (2) The bright spot disturbs one quadrant of the disc, so that one cannot, for instance, assume that temperature varies only with radius. (3) If the discs are flared, rather than flat, they will have a cool 'wall' at their outer rim. This can partially obscure the inner disc, particularly at high inclinations — but a high inclination is required to see eclipses! In principle these effects can be accounted for in the mapping, but their presence reduces the objectivity of the technique since the result will depend on how they have been dealt with. There are, though, bonuses as well: for instance, the quadrant of the eclipse map affected by the bright spot can yield the bright-spot temperature.

40 Spectral characteristics

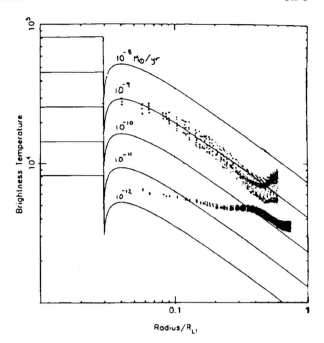

Fig. 3.7: The temperature (in Kelvin) at different disc radii in Z Cha during two observations, derived from eclipse mapping (see Box 3.2). In the upper data the disc is brighter and follows the model curve well; on the other occasion the disc is fainter and has a flatter temperature distribution. The model curves are drawn for several different mass-transfer rates (note the $T \propto r^{-3/4}$ behaviour at large radii and the reduction in temperature approaching the white dwarf). (Figure by Keith Horne.[7])

are emitted when electrons jump between atomic energy levels that are determined by fundamental physics. However, consider an atom emitting a photon when moving towards us. It emits one wavecrest towards us and then, after a set amount of time, the next wavecrest. But since it has moved during that time, the second wavecrest will be nearer the first wavecrest than it would have been were the atom stationary. Thus the distance between successive crests is reduced, and the photon effectively has a shorter wavelength. This is called a *blue shift*, since an optical photon would appear shifted towards the blue end of the spectrum. Similarly, motion away from us increases the separation between wavecrests and produces a *red shift*.

The amount of the Doppler shift is directly proportional to the speed at which the material is moving along our line of sight.[‡] This is invaluable, as it means that we have only to measure the wavelength of a spectral line (easily done with a spectrograph) to know one component of the motion of the emitting material (unfortunately there is no similarly easy way of deducing the motion across our line of sight, in the plane of the sky).

3.4.1 S-waves

Since most parts of a cataclysmic variable are moving at a few hundred to a few thousand $\mathrm{km\,s^{-1}}$, the lines are typically shifted by a few to tens of Ångstroms, an amount which is readily detectable.

[‡]For speeds much less than the speed of light, $v \ll c$, which always applies in cataclysmic variables, the shift $\Delta\lambda$ is given by $\Delta\lambda/\lambda = v/c$ where λ is the wavelength at rest.

Consider, firstly, the case of a small region circling with the orbital motion — the bright spot is often a good example — in a binary that is edge-on. At one orbital phase it will head straight towards us, producing blue-shifted lines, while half an orbit later it will produce red-shifted lines. If its orbit is circular the shift will change sinusoidally from blue to red and back. Thus if a series of spectra taken over an orbital cycle are plotted in sequence, the line feature will snake back and forth in a characteristic wave that is referred to, particularly when produced by the bright spot, as an *S-wave* (see Fig. 3.8).

The maximum red or blue shift of an S-wave in an edge-on binary reveals the actual speed of the material. More generally, for binaries at an inclination i, motion v in the binary plane can be resolved into a component $v \sin i$ along our line of sight and a component across our line of sight. Thus at lower inclinations the amplitude of the S-wave is reduced by $\sin i$. As an example of how this is used, it is often found that the amplitude of the S-wave from a bright spot reflects not the velocity of the free-falling accretion stream, nor the Keplerian velocity of the disc material that it has hit, but a velocity midway between the two.[8] This suggests that the line emission from the bright spot results from the energy being dissipated as the stream material is entrained into the disc.

Although the disc and bright spot generally dominate the line emission, in some cataclysmics one can detect emission or absorption lines from the red and white dwarfs. These also show S-wave behaviour owing to the stars' orbital motion about the common centre of mass of the binary. The mean velocity of the line averaged over the whole orbit gives the (line-of-sight) velocity of the binary itself with respect to Earth.

3.4.2 Double-peaked lines from the accretion disc

A disc can be considered as a collection of small regions emitting S-waves. Its line profile is thus a sum of S-waves over a range of amplitudes (since velocity decreases as radius increases), weighted by the differing area of the disc at different radii. Since there is less area at small radii, the line peters out in the high-velocity wings corresponding to the innermost orbits. Conversely, the increase in area allows the outer disc to dominate the emission. The result (see Fig. 3.9) is a double-peaked profile in which the peaks are shifted from the line centre by a velocity typical of the outer disc, though again reduced by the $\sin i$ projection factor.

The disc is, of course, centred on the white dwarf, and so follows its motion about the common centre of mass. This means that the whole double-peaked profile of the disc executes an S-wave in sympathy with the white dwarf.

If, though, one measures the separation of the double peaks one normally deduces an outer-disc velocity that is below the expected Keplerian velocity by 10–30%. The reasons for this are unclear, but could involve the effect of the secondary, distorting the outer-disc orbits, or influences on the line profile formation that we have not taken into account. In fact, while many cataclysmics show the double-peaked line profiles expected from a disc, they still differ in detail from the theoretical profile. Furthermore, some systems do not show double peaks at all. Indeed, some

Fig. 3.8: The profile of the Hα line of WZ Sge changes over the orbital cycle. The line profile at each phase is displayed as a greyscale, with darker colouring indicating greater intensity. The data show the double peaks characteristic of an accretion disc, with an 'S-wave' running from side to side over the orbit. Note also a reduction in emission near phase 1.0, owing to an eclipse. The approaching wing of the disc (and hence the blue-shifted emission) is eclipsed first, followed by the receding wing, creating a 'rotational disturbance'. (Data by Henk Spruit and Rene Rutten.[9])

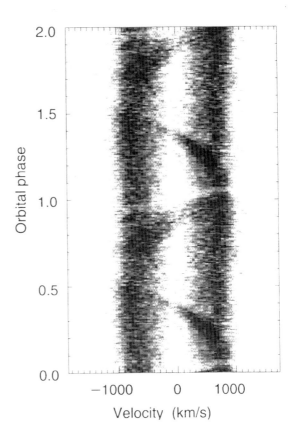

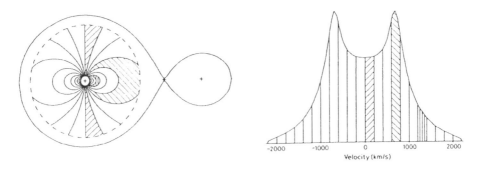

Fig. 3.9: *Left:* A Keplerian disc with contours of the different velocities as projected onto the line of sight (the viewer is below the plot). *Right:* The resulting double-peaked profile. The different shadings match corresponding regions in the two plots. (Figures by Keith Horne and Tom Marsh.[10])

high-inclination eclipsing systems, in which double-peaks are expected to be most obvious, show only single-peaked lines. Thus, line formation in accretion discs is only partially understood.

During eclipse the approaching wing of the disc is eclipsed first (see Fig. 2.10), so the blue-shifted line peak disappears just before mid-eclipse, followed by the red-shifted line peak just after mid-eclipse. This characteristic pattern, called a *rotational disturbance*, is the most direct indication of the presence of a circulating accretion disc in cataclysmic variables.

Box 3.3: Doppler tomography

Doppler tomography is a technique for presenting spectra of cataclysmic variables that has been popularised by Tom Marsh and Keith Horne.[11] The rationale is that while the practised eye can happily trace S-wave features in a sequence of spectra, humans respond more readily to pictures. Unfortunately, translating the line emission into a picture of a disc would require knowledge of the velocity at each point in the disc, which we do not have without making assumptions of dubious validity. We can, though, take a half-way step, and create a map of the velocities within the disc, without translating these into positions.

The orbital motion of a small region of disc is specified by a velocity and a phase. This can be plotted on a polar diagram (r, θ) where the vector length r represents the speed of the motion and the angle θ from the vertical represents the phase (360° corresponding to one cycle). Each location on such a diagram produces an S-wave — indeed if one rotated the diagram about its origin, the projection of the feature onto the (stationary) x-axis would trace out the S-wave.

The essence of tomography is to perform the reverse process, using a set of spectra taken over the orbital cycle. For each location in the velocity plot, one adds up the flux along the track of the corresponding S-wave, and shades the location in proportion to the total flux. The result is a tomogram such as that in Fig. 3.10.

In a tomogram the secondary, being closest to us at phase 1, is at the top; similarly the primary is also on the y-axis, shifted below the origin by its orbital velocity. The disc in a tomogram is still circular, and centred on the white dwarf's velocity, but is 'inside-out', the low-velocity (outer) regions being nearer the origin. If the orbital period, mass ratio and inclination of the system are known, then the Roche geometry, stellar velocities, and the velocities along the stream can be calculated and plotted on the tomogram, as an aid to interpretation. If, though, the phasing of the orbit is uncertain, as is often the case in non-eclipsing systems, any phase error leads to a rotation of the tomogram.

There are two assumptions of tomography that can cause problems. First, it assumes that all components have the same mean velocity with respect to Earth. This will be valid for anything orbiting in the binary plane, but is violated by material moving out of the plane. Emission from such material will appear 'defocused' in the tomogram. More seriously, tomography assumes that a component is equally visible at all orbital phases. Thus occultation or optical-depth effects which change the visibility of components can produce spurious features in the tomogram.

44 Spectral characteristics

Overall, the advantage of Doppler tomography is that considering spectra in a different way stimulates the thought processes when interpreting observations. However, a velocity plot is hardly more intuitive than a time-resolved sequence of spectra. It is also possible for tomography to introduce artifacts, so one should be sceptical of any interpretation that cannot be sustained by inspection of the trailed spectra themselves.

Perhaps the true potential of tomography will come from combining it with eclipse mapping, so that the spatial information from the eclipse combines with velocity information from the tomograms to produce a genuine picture of a disc. This, though, is still under development, primarily because it would require spectacularly good data, with high-quality spectral profiles obtained every few seconds to match the time resolution used in eclipse mapping.

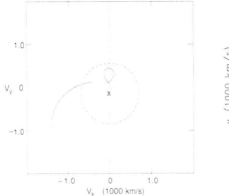
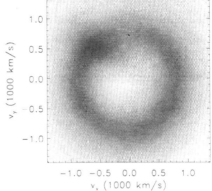

Fig. 3.10: A schematic tomogram (*left*) showing the secondary with the stream flowing from its tip. The cross marks the white-dwarf velocity, and the dashed circle illustrates disc emission. The axes are the velocities projected onto the x and y directions. At *right* is a tomogram of the WZ Sge line profiles from Fig. 3.8 (by Henk Spruit[9]), showing a ring of disc emission and a bright spot from the stream/disc collision (which has a velocity midway between that of the stream and the local disc).

3.5 DERIVING MASSES AND OTHER PARAMETERS

We now have all the ingredients necessary to deduce the basic properties of a cataclysmic variable, such as the stellar masses. While such studies are the foundation of quantitative cataclysmic variable research, one can follow the rest of this book without knowing the details of how this is accomplished. For this reason the material is placed in Appendix A.

Chapter 4

The evolution of cataclysmic variables

When I introduced white dwarfs in Chapter 2 I explained that they form as the core of a red giant, whose fluffy outer layers extend to ~ 100 solar radii (when the Sun becomes a red giant it will engulf the Earth). I also explained that the orbital periods of cataclysmic variables imply that the separation of the two stars is only ~ 1 solar radius. So what happened before the white dwarf formed; was the secondary orbiting *inside* a red giant?

4.1 THE ORIGIN OF CATACLYSMIC VARIABLES

Stars form when a cloud of interstellar dust and gas collapses under its own gravity. Large clouds containing thousands of solar masses of material are the most likely to collapse, forming a whole cluster of young stars. Thus stars form near companions, and find themselves gravitationally bound into binaries, triples, pairs of binaries, or similar combinations — only a minority of stars are single like our Sun.

The stars destined to become cataclysmic variables begin as binaries separated by a few hundred solar radii, orbiting every ~ 10 years; one must be less than a solar mass, the other more massive. The heavier star evolves more rapidly, since the greater weight on its core ensures a higher pressure and temperature, and thus more vigorous nuclear reactions.* Eventually the massive star expands to become a red giant; it then overflows its Roche lobe, transferring its outer layers to the lower-mass companion.

But this situation — the reverse of that in a cataclysmic variable — is unstable. The heavier star is nearer the centre of mass of the binary, so material transferred to its companion moves further from the centre of mass. This increases the angular momentum of the transferred material, and thus the stellar separation decreases slightly, so that angular momentum is conserved overall. But the decrease in separation decreases the Roche-lobe size; the heavier star finds itself overfilling the

---

*Luminosity scales $\propto M^3$ but fuel reserves scale $\propto M$, thus the lifetime is $\propto 1/M^2$.

Roche lobe even more, and yet more material is transferred. The result is a runaway feedback as the entire envelope of the red giant is dumped onto the companion star, limited only by the speed at which the material can flow. This might take only a few years, which is extraordinarily fast by the standards of stellar evolution![†]

The companion star cannot assimilate such an influx, so the material overfills both Roche lobes, and forms a cloud surrounding the two stars. In this 'common envelope' phase the nascent cataclysmic variable is effectively orbiting within a red giant. The effect is like swimming in treacle: the drag on the stars as they orbit drains their orbital energy, causing them to spiral inwards. Their separation shrinks from ~ 100 $R_\odot$ to ~ 1 $R_\odot$ in about 1000 years.

From the point of view of the envelope, the binary acts as a propeller, expelling it outwards. The energy extracted from the binary orbit pushes the envelope into interstellar space, forming a 'planetary nebula'.[‡] The now-naked binary is either a cataclysmic binary, or, if the separation is still too large for mass transfer, a detached red-dwarf/white-dwarf binary.

4.2 DRIVING MASS TRANSFER

As discussed above, transferring mass from a heavier star to a lighter companion is unstable and leads to rapid, catastrophic mass exchange. Now consider the opposite situation (as in cataclysmics) of material transferring from a lower-mass secondary onto a white dwarf, again assuming that angular momentum is conserved. The lower-mass star is further from the centre of mass, so transferred material ends up closer to the centre of mass, losing angular momentum, and hence the binary separation increases slightly to compensate. The increase in separation causes the secondary to detach from its Roche lobe, halting further mass loss (this is discussed mathematically in Box 4.1).

Under what situations can the steady, long-lived mass transfer of a cataclysmic variable occur? One possibility involves a secondary that is evolving into a red giant. The expansion of the secondary keeps it in contact with its Roche lobe and enables steady mass transfer. However, this cannot explain most cataclysmic variables, since they contain secondary stars of less than a solar mass, whose lifetimes are very long. The Universe is not yet old enough for such stars to have evolved into red giants.

The other possibility is a gradual loss of angular momentum from the binary. This shrinks the orbit, and hence the secondary's Roche lobe, allowing the transfer of material excess to the new size of the Roche lobe. We think that there are two main mechanisms by which cataclysmic binaries lose angular momentum: gravitational radiation and magnetic braking.

[†]We revisit this issue in Section 11.6.
[‡]This confusing term arose because such nebulae look like planets in small telescopes, but otherwise they have nothing to do with planets.

4.2.1 Gravitational radiation

In accordance with the theory of general relativity, matter causes space to curve, although we cannot perceive this warping because we cannot see in four dimensions. The repetitive orbiting of two stars causes a rhythmic warping of space which ripples outwards in a periodic wave. We call this wave *gravitational radiation*. The energy to generate the wave is extracted from the binary orbit, causing a slow spiralling inwards.

Unless one is near a neutron star or black hole, relativistic effects are comparatively weak, so gravitational radiation is negligible for most binaries. But as binaries orbit closer and closer their orbital speeds increase, boosting the gravitational radiation,[§] and it becomes significant in the shortest-period systems.

The amount of gravitational radiation can be calculated from relativity theory, and, for a cataclysmic binary with a 2-hr orbital period, drives mass transfer at a rate of $\sim 10^{13}$ kg s^{-1} (which is $\sim 10^{-10}$ M$_\odot$ yr^{-1}). The concept of the mass-transfer rate will occur often in the remainder of the book, and I will denote it by the mathematical shorthand $\dot{M}$ (pronounced m-dot).

4.2.2 Magnetic braking

The ingredients of *magnetic braking* are a stellar wind and a stellar magnetic field. A stellar wind is the stream of energetic ionized particles (mostly electrons and protons) ejected from a star. We detect such particles from the Sun — they cause the aurora when they hit the Earth's atmosphere — and expect similar winds from the secondaries in cataclysmics.

We also observe many stars to have strong magnetic fields, though how they are produced — perhaps by dynamo action deep within the star, where convection forces bubbles of gas into circular motions — is poorly understood. It appears that rapid rotation of a star generates stronger fields, which implies that red dwarfs in cataclysmics should be highly magnetic (tidal forces ensure that they corotate with the orbital motion, giving rotation periods of hours, whereas single stars of a similar type would take days to rotate).

Electromagnetic theory tells us that the particles in the stellar wind, which possess an electric charge, cannot easily cross magnetic field lines and instead flow along them. Thus they are forced to corotate with the red dwarf and its magnetic field. Like a pebble out of a slingshot, they are accelerated to high speeds and then shot off into space. The long lever-arm of the field lines ensures that they take with them substantial angular momentum.

The effect of the angular-momentum drain is to brake the rotation of the red dwarf. However, as just stated, the rotation is locked to the binary orbit by tidal interactions, which ensure that the red dwarf always points the same face to the white dwarf (for the same reason the Moon always points the same face to Earth). Ultimately, therefore, the angular momentum is supplied by the orbit, which shrinks in consequence.

[§]Angular-momentum loss rate is given by $\dot{J}/J \propto M_1 M_2 M/a^4$.

Box 4.1: The binary's response to mass transfer

In this Box I will repeat the discussion of mass transfer in Section 4.2 more rigorously. We first need the angular momentum of the binary. From the basic formula $J = mrv$, where the velocity v is perpendicular to the lever arm r, we can write (see Appendix A)

$$J = M_1 a_1 \frac{2\pi a_1}{P_{\text{orb}}} + M_2 a_2 \frac{2\pi a_2}{P_{\text{orb}}}$$

which combined with $a = a_1 + a_2$ and $a_1 M_1 = a_2 M_2$, and eliminating P_{orb} using Kepler's law (Box 2.1), leads to

$$J = M_1 M_2 \left(\frac{Ga}{M}\right)^{1/2}$$

where $M = M_1 + M_2$. Logarithmic differentiation (that is, taking natural logs and differentiating with respect to time) then gives

$$\frac{\dot{J}}{J} = \frac{\dot{M}_1}{M_1} + \frac{\dot{M}_2}{M_2} - \frac{1}{2}\frac{\dot{M}}{M} + \frac{1}{2}\frac{\dot{a}}{a}.$$

If the total mass is conserved ($\dot{M} = 0$) then

$$\frac{\dot{a}}{a} = 2\frac{\dot{J}}{J} + 2\frac{-\dot{M}_2}{M_2}\left(1 - \frac{M_2}{M_1}\right).$$

Thus if angular momentum is also conserved ($\dot{J} = 0$), transfer from the secondary (for which $-\dot{M}_2$ is positive) leads to increasing a (positive $\dot{a}$) provided $M_2 < M_1$. Further, using the expression for the secondary's Roche lobe (Box 2.4)

$$R_2 \propto a \left(\frac{q}{1+q}\right)^{1/3} = a \left(\frac{M_2}{M}\right)^{1/3}$$

and logarithmically differentiating, we have

$$\frac{\dot{R}_2}{R_2} = \frac{\dot{a}}{a} + \frac{1}{3}\frac{\dot{M}_2}{M_2} - \frac{1}{3}\frac{\dot{M}}{M}$$

which combines with the previous expression (and again setting $\dot{M} = 0$) to give

$$\frac{\dot{R}_2}{R_2} = 2\frac{\dot{J}}{J} + 2\frac{-\dot{M}_2}{M_2}\left(\frac{5}{6} - \frac{M_2}{M_1}\right).$$

Thus if $\dot{J} = 0$, mass transfer causes the secondary's Roche lobe to expand ($\dot{R}_2$ positive) provided $q < 5/6$. Sustained and stable mass transfer (requiring $\dot{R}_2$ negative) is only possible through angular momentum loss ($\dot{J}$ negative).

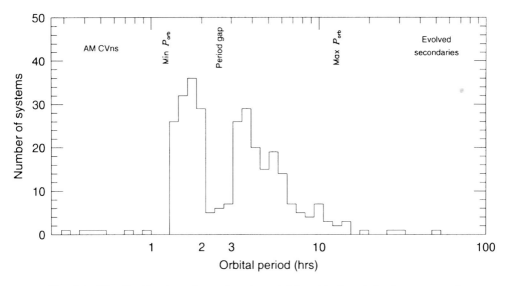

Fig. 4.1: The distribution of cataclysmic variable orbital periods, from a compilation by Hans Ritter and Uli Kolb.[1]

The theory of magnetic braking is too uncertain to yield the rate at which angular momentum is lost, and hence the rate of mass transfer, so we have to deduce it observationally. If we know the white dwarf mass, we can add up the emitted luminosity to deduce the rate at which material accretes onto the white dwarf (see Box 4.2). Of course, to determine the luminosity from the observed flux we need to know how far away the system is, and such distances are uncertain. Another difficulty is that much of luminosity emerges in the UV region of the spectrum, which is hard to observe. Nevertheless, we estimate that magnetic braking causes a mass-transfer rate, $\dot{M}$, of $\sim 10^{14}$ kg s^{-1} (which is $\sim 10^{-9}$ M$_\odot$ yr^{-1}, so that a tenth of a solar mass is transfered in 100 million years). Individual systems, though, can have transfer rates differing from this by factors of 10–100, for reasons that are only poorly understood (see Chapter 12).

4.3 THE DISTRIBUTION OF ORBITAL PERIODS

We now have the ingredients needed to understand the evolution of cataclysmic variables. Of course we cannot observe this directly, the timescales being of order 10 million years. We can only observe the population of binaries as they are now and try to deduce their relative ages, theorising about whether one system will evolve to become like another.[2]

The starting point is the set of orbital periods, now known for 300 different cataclysmics, hard won by hundreds of hours of observation. Their distribution is shown in Fig. 4.1.

> **Box 4.2: Estimating distances and mass-transfer rates**

By conservation of energy, the total luminosity L of a cataclysmic variable in equilibrium equals the released gravitational energy. For a white-dwarf mass M_1, creating a gravitational well of potential $U = -GM_1 m/r$, the rate of gravitational energy release for a flow $\dot{M}$ onto the white-dwarf surface at R_1 is

$$|\dot{U}| = \frac{GM_1 \dot{M}}{R_1} = L$$

where M_1 and R_1 can be estimated using the methods of Appendix A.

The flux (energy m^{-2}) observed at Earth is $L/4\pi D^2$, where D is the distance to the binary. Thus we can estimate L and hence $\dot{M}$ if we (1) add up the observed flux over the whole spectrum, including the hard-to-observe extreme UV where the energy of a cataclysmic variable peaks, and (2) estimate D.

The most reliable estimate of a star's distance is gained from its parallax. This is the slight wobble of nearby objects against background objects as the viewing position shifts from side to side. The orbit of the Earth around the Sun shifts our view of nearby stars sufficiently to detect the wobble for stars out to 100–200 parsecs (the unit parsec, abbreviated pc, is the distance a star would be if it showed a parallax of one arcsec; it equates to 3.26 light years or 3.09×10^{16} m). However, most cataclysmics are too distant and faint to give accurate parallaxes, which currently requires observations with the *Hubble Space Telescope*. At the time of writing these are available only for SS Cyg (giving a distance of 166 ± 13 pc), U Gem (96.4 ± 4.6 pc) and SS Aur (200 ± 26 pc).[3] Distances for some other objects, using the methods outlined below, are given in Appendix F.

The commonest method of estimating distances is to compare the light from the secondary star with that expected from a similar single star. One first finds distances and radii for single red dwarfs, perhaps by using the parallax method on nearby examples, and so computes their brightness per surface area at a given wavelength. One then picks the red dwarf nearest in type to the secondary star, by matching the spectral features of each. Assuming the secondary to be unaffected by being in a binary, it will have the same surface brightness as the single star. Hence, knowing the surface area of the Roche lobe from the orbital period (it also depends weakly on q), one can compute the luminosity of the secondary star at the chosen wavelength. By comparing this with the observed luminosity, and applying the inverse square law, the distance to the binary can be determined.

As ever there are potential pitfalls, the main ones being: (1) possible bias in the calibration of single stars; (2) difficulty in distinguishing the light of the secondary from disc light (one usually observes in the IR to minimise this problem); and (3) distortion of the secondary's spectrum by irradiation (one can watch how features vary round the orbit to judge the effect of this). Given these difficulties, unreliable distances are usually the biggest source of uncertainty in the estimates of luminosity and mass-transfer rate.

4.3.1 The long-period cutoff

Notice, first, that the number of systems dwindles for orbital periods above ~ 12 hrs. This results from the requirement that the secondary be less massive than the white dwarf to avoid rapid and catastrophic mass transfer. Since the white dwarf must be below the Chandrasekhar limit of ≈ 1.4 $M_\odot$ (see Section 2.1) the red dwarf must be also. The size of the binary (and hence the secondary's Roche lobe) increases with orbital period. Filling a larger Roche lobe requires a heavier star, and so the secondary mass increases with orbital period (see Appendix A.5). The mass limit thus leads to a limit on the orbital period of ~ 12 hrs. Few white dwarfs, however, are as heavy as the Chandrasekhar limit, so the $q < 1$ requirement begins to reduce the number of cataclysmics from ~ 6 hrs upwards.¶

The occasional system with a much longer period (for example, GK Per has an orbital period of 48 hrs) is explained as containing a secondary on its way to becoming a red giant, making it bigger but less massive than a normal star. Since the secondary is expanding in such systems, they evolve to longer periods.

The period at which a binary becomes a cataclysmic variable will depend on the size of the red dwarf when the binary emerges from the common envelope phase. If it is smaller than its Roche lobe then no mass transfer occurs, and the system is often called a 'pre-cataclysmic variable'. In such systems magnetic braking decreases the binary separation, over hundreds of millions of years, until the Roche lobe makes contact with the red dwarf, initiating mass transfer. Thus systems with lower-mass secondaries become cataclysmics at shorter orbital periods. Once in contact, magnetic braking continues the evolution to shorter periods, and the steady loss of material causes the red dwarf to shrink along with its Roche lobe.

4.3.2 The period gap

Magnetic braking can explain a steady evolution to shorter periods, but when they arrive at ~ 3 hrs something appears to happen: there is an abrupt drop in the number of systems in the range 2–3 hrs, referred to as the *period gap*. Below this range ($P_{\rm orb} < 2$ hrs) systems have mass-transfer rates characteristic of gravitational radiation ($\dot{M} \sim 10^{-10}$ $M_\odot$ yr^{-1}) rather than the higher rates characteristic of magnetic braking ($10^{-9} - 10^{-8}$ $M_\odot$ yr^{-1}) seen above the gap ($P_{\rm orb} > 3$ hrs).

The standard explanation suggests that magnetic braking switches off when a cataclysmic binary has evolved down to 3 hrs. But why, then, do we see a gap, rather than simply a change to a lower $\dot{M}$? This is because the secondary has been driven out of equilibrium by the mass transfer. As the secondary loses its outer layers, the weight on the core decreases, and so nuclear reactions in the core decrease. With less energy being generated in the core, the pressure drops, and the star contracts slightly under gravity. But the contraction takes place on the

---

¶For some reason mass transfer from a lighter companion is counterintuitive, and many illustrations show (incorrectly) a secondary whose Roche lobe is larger than the white dwarf's Roche lobe. This is particularly prevalent in general textbooks which describe cataclysmic variables briefly, and so is a good test of whether the author has researched his subject properly!

thermal timescale.^|| If this timescale is longer than the timescale on which material is transferred — as can occur during magnetic braking — the star cannot adjust to mass transfer rapidly enough, and finds itself with too large a radius for its mass.

When magnetic braking shuts off, the secondary contracts to its correct radius, detaching from its Roche lobe. To resume mass transfer the orbit must decrease, shrinking the Roche lobe back towards contact. Gravitational radiation effects this reduction, and at a period of ~ 2 hrs contact is re-established and mass transfer resumes. From here on to shorter periods, mass transfer and evolution proceed at the lower rate driven by gravitational radiation.

Hence, while evolving from 3 hrs down to 2 hrs, cataclysmics are detached, and, with no mass transfer occurring, are too faint to be seen. An exception can occur if a cataclysmic is born (first achieves contact) with a period of 2–3 hrs. Since it has not been driven out of equilibrium it can simply transfer matter under gravitational radiation; this explains the few systems found in the gap. [The standard picture is altered if the white dwarf posseses a strong magnetic field (see Chapter 8) which can couple to the field of the secondary and affect the evolution. Thus magnetic systems can also populate the gap.]

For the explanation outlined above to be satisfactory we need a reason for magnetic braking to become ineffective when a cataclysmic variable reaches a period of 3 hrs. In principle this could occur if the secondary's magnetic field becomes weaker, if the field configuration changes so that few open field lines connect to the interstellar medium, or if the stellar wind turns off. One clue is that at the mass appropriate to a $\sim$ 3-hr binary, the convective motions in a red dwarf circulate throughout the star, removing the non-convective core present at higher masses. This might quench the magnetic dynamo. Arguing against this is the fact that signs of magnetic activity are still observed in lower-mass single stars. However, single stars do not rotate as rapidly as those in close binaries, and they are not affected by the tidal forces that act in binaries, so this argument is not conclusive. Perhaps when the time taken to rotate (which is locked to the orbit and so decreases as cataclysmics evolve) becomes shorter than the timescale of convective motions, the dynamo switches off. Overall, we have a poor understanding of magnetic dynamos in stars, and without this we are only guessing at the origin of the cataclysmic-variable period gap. Note that there is observational evidence that some stars below the period gap transfer matter at a rate greater than that driven by gravitational radiation (see Section 12.5). Thus, some residual magnetic braking may also be important below the gap.

4.3.3 The period minimum

The third obvious feature of the distribution of orbital periods (Fig. 4.1) is their sudden cutoff at a minimum period of ~ 78 mins. This occurs when the mass of the red dwarf becomes so low that it starts behaving like a white dwarf, becoming *degenerate*. As explained in Section 2.1, such stars are supported not by the pressure

---

^|| Also called the Kelvin–Helmholtz timescale, this is the time for stored energy to leak away, and so is the ratio of total energy to luminosity. For a red dwarf in a 3-hr binary it is $\sim 5 \times 10^8$ yrs.

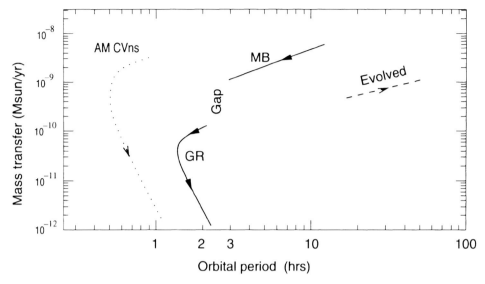

Fig. 4.2: The evolution of cataclysmic variables proceeds by magnetic braking (MB) above the gap, and then by gravitational radiation (GR) below the gap and through the period minimum. The helium systems (AM CVns) evolve on a similar track at shorter periods. Systems with evolved secondaries are at longer periods. Arrows show the direction of evolution.

of the gas, but by the quantum-mechanical requirement that adjacent atoms cannot be too close. This leads to a peculiar behaviour of their radii. Whereas normal stars are larger when they are more massive, white dwarfs are *smaller* when they are more massive. Thus when mass transfer reduces the mass of a degenerate secondary, it responds by expanding!

To appreciate the effect this has on the evolution of a cataclysmic variable, it is easiest to consider mass transfer as a two-stage process: a blob of accretion followed by the response of the binary (though in reality the two operate continually). In a system with a normal secondary the response to a blob of mass transfer is that the binary expands and the Roche lobe detaches (see Section 4.2), while the secondary radius is reduced slightly. Angular-momentum loss then reduces the separation and Roche lobe until contact is resumed at the shorter period appropriate to the smaller secondary. Thus the system evolves to shorter periods.

With a degenerate secondary, the binary still expands and the Roche lobe detaches after a blob of accretion, but the secondary also expands slightly. When brought back into contact the period will be slightly longer, appropriate to the now-larger secondary. Thus the system evolves to longer periods.

Detailed modelling of the above process (which involves the mass-radius relation of the secondary as it becomes degenerate, and the fact that the secondary is driven out of equilibrium by mass transfer) shows that cataclysmics pass through a minimum period of about 78 mins and then evolve to longer periods. This is in good agreement with what we actually observe.

4.3.4 The ultimate fate

At the 78-min minimum period the mass of the secondary has dropped to only ≈ 0.06 $M_\odot$, and continues to decrease as the binary period lengthens again. By the time the period reaches 100 mins it is only ≈ 0.02 $M_\odot$. The low-mass secondary can no longer make much of a ripple in space as it orbits, and this, coupled with the lengthening period, means that gravitational radiation decreases rapidly. Evolution slows down, the mass-transfer rate plummets (to only $\approx 4 \times 10^{-12}$ $M_\odot$ yr^{-1} at P_{orb} = 100 mins) and the binaries become faint and hard to detect. Ultimately, the secondary star reduces to the mass of a planet, and the end result is a Jupiter-like object orbiting a white dwarf.

We have no proof that any of the known cataclysmics have evolved past the period minimum, but there is speculation that WZ Sge, with an orbital period of 82 mins and a very low accretion rate of only $\sim 10^{-11}$ $M_\odot$ yr^{-1}, may be such an object.

4.4 AM CVN STARS

Although the number of cataclysmic variables drops dramatically at the 78-min period minimum, a few systems have even shorter periods (Fig. 4.1). Such systems have secondaries composed mostly of helium (whereas in normal stars 92% of the atoms are hydrogen and only 8% helium). Since helium is heavier than hydrogen, helium stars are more compact for the same mass, and so are in contact with their Roche lobes only at smaller orbital separations. Their evolutionary track parallels that for hydrogen secondaries, but shifted to shorter periods. Gravitational radiation is enhanced by the smaller orbits, driving a higher mass transfer, though this again drops as the secondary loses mass, becomes degenerate, and passes through a period minimum.

The helium-rich cataclysmics, known as AM CVn stars after the first to be found, are relatively rare since their existence depends on particular conditions in the initial binary. One route to an AM CVn star is a binary in which both stars turn into red giants, one after the other; the second common-envelope phase could then reduce the separation of the two helium cores sufficiently to allow mass transfer.

AM CVn stars can be thought of as compact versions of cataclysmic variables, but they contain accretion discs of nearly pure helium. Comparing their properties with those of hydrogen-rich discs can probe and test our understanding of disc behaviour.

Chapter 5

Discs and outbursts

Of all the phenomena displayed by cataclysmic variables, the most obvious, the most characteristic, and the one giving rise to the name 'cataclysmic', is the outburst. Referred to as *dwarf nova* outbursts — to distinguish them from the even more dramatic but much rarer nova eruptions (see Chapter 11) — the intensity of many cataclysmics jumps by several magnitudes in the space of a day or so, to stay bright for about a week before declining. After some months of quiescence the outburst repeats.

5.1 DWARF NOVA OUTBURSTS

Outbursts of U Gem have been followed since their discovery in 1855, and are illustrated in Fig. 5.1. They are semi-regular, but the outburst durations and the recurrence times are variable. Furthermore, the lightcurve is characteristic of the star, as can be seen by comparing U Gem with SS Cyg (Figs 1.1 and 5.1). Whereas outbursts of U Gem repeat, on average, every 100 days, SS Cyg repeats in half that time, though in both stars outbursts can be ~ 20 days early or late. This unpredictability means that discovering the nature of dwarf novae requires sustained monitoring — an invaluable service rendered by amateurs worldwide.

In jumping from 14th to 9th magnitude, U Gem brightens by a factor of 100 in only a day, which led the early observers to conclude that some cataclysm had enveloped the star. But although outbursts have been observed for well over a century, it is only recently that we have come to understand them. The first clues required the development of high-speed photometers (pioneered in the 1970s by Brian Warner and Ed Nather[1]) capable of recording high-quality lightcurves during outbursts. As ever, eclipsing systems yield the most information. Fig. 5.2 records the lightcurve of OY Car caught on the rise to outburst. As the intensity increases, the eclipse profile changes from one dominated by the white dwarf and bright spot to one dominated solely by the disc component (Fig. 5.3). Thus the system is in outburst because the disc has brightened dramatically.

In 1974, Yoji Osaki suggested a cause for the outburst, based on an instability in the accretion disc.[2] He reasoned that if material was transferred from the secondary

56 Discs and outbursts

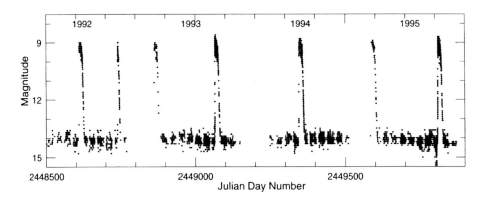

Fig. 5.1: A 3-yr section of U Gem's lightcurve showing semi-regular 5-mag outbursts. Julian Day Number is explained in Appendix C. [Data compilation by the American Association of Variable Star Observers (AAVSO).]

star at a constant rate, and if this rate was higher than could be transported through the disc by viscous interactions, then material would pile up in the disc. Eventually, the pile-up might cause the disc to become unstable, boosting the viscosity, greatly increasing angular-momentum transport, and so spreading the excess material both inwards onto the white dwarf and outwards. The increased accretion onto the white dwarf both enhances the luminosity of the system and drains the disc of its matter. Soon the disc drops back into a quiescent, low-viscosity state, where it it gradually replenished by the mass-transfer stream, building it up towards a new outburst.

For many years an argument raged between Osaki's model of a disc instability

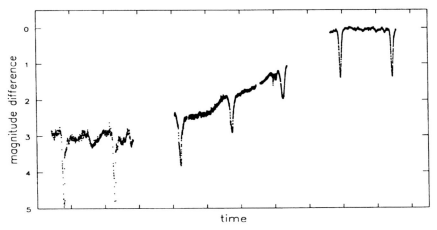

Fig. 5.2: Orbital lightcurves of the eclipsing system OY Car in quiescence, on the rise to outburst, and at outburst maximum. The three sections of data were recorded on different nights. (Figure by Nicholas Vogt and Rene Rutten.[3])

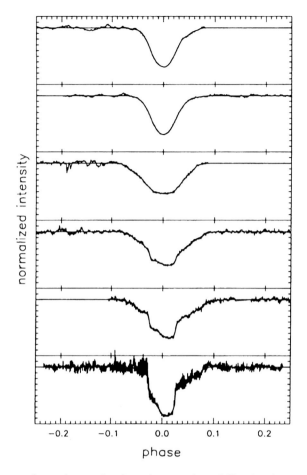

Fig. 5.3: The eclipses from Fig. 5.2 stacked from quiescent (bottom) to brightest (top). The eclipse profile in quiescence is dominated by a bright white dwarf and bright spot (compare to Fig. 2.12), whereas in outburst the disc brightens to outshine everything, so that the eclipse has a smooth profile dominated solely by the disc eclipse. (Figure by Rene Rutten and Nicholas Vogt.[3])

and an alternative based on an instability in the secondary star. Developed principally by Geoff Bath, this model suggests that the secondary spouts bursts of mass transfer which flood into the disc and raise its temperature and luminosity.

Opinion has settled in favour of Osaki's disc-instability model for two reasons: firstly, several observational results support it, and, secondly, in the 1980s a convincing theoretical model for the instability was developed.

The chief difficulty for the mass-transfer-burst model was the observation that the luminosity of the bright spot stayed roughly constant during outburst, whereas one would expect its luminosity to increase in proportion to the greatly enhanced flow of mass though it. Indeed the bright-spot luminosity strongly supports the disc-instability model (see Box 5.1).

Another problem relates to the changes in the radius of the disc over the outburst cycle. In the disc-instability model the increased viscosity spreads material both outwards and inwards, so that the disc expands at the start of outburst. Then, during quiescence, the addition of more material from the stream causes the disc to shrink gradually until the next outburst occurs (this is because the stream material

58 Discs and outbursts

has the specific angular momentum appropriate to the circularisation radius, which is less than the specific angular momentum at the disc edge, and the addition of lower-angular-momentum material causes the disc to shrink).

In contrast, the mass-transfer burst of the competing model would cause the disc to shrink at the onset of outburst, for the reason just stated, piling material up at the circularisation radius. In quiescence this material would spread viscously, gradually expanding the disc. The observations clearly show that the disc contracts during quiescence and expands during outburst (see Fig. 5.4), thus favouring the disc-instability model. This, together with the success of its theoretical underpinnings, have led to its near-universal acceptance. However, it is still possible that an increase

---

Box 5.1: Osaki's argument for a disc instability

The main argument put forward by Osaki in support of a disc-instability explanation for outbursts compares the luminosity of the bright spot in quiescence to the luminosity of the outburst.[2] Steady mass transfer from the secondary at a rate $\dot{M}_{sec}$ will hit the bright spot and release gravitational energy at a rate

$$L_{spot} = \frac{GM_{wd}}{R_{disc}} \dot{M}_{sec}$$

where R_{disc} is the radius of the disc edge. Further, accretion onto the white dwarf during outburst at a rate $\dot{M}_{out}$ releases energy

$$L_{out} = \frac{GM_{wd}}{R_{wd}} \dot{M}_{out}.$$

For U Gem Osaki estimated $R_{disc}/R_{wd} \approx 32$ to obtain

$$\frac{L_{out}}{L_{spot}} \approx 32 \frac{\dot{M}_{out}}{\dot{M}_{sec}}.$$

Now, if the mass transferred from the secondary accumulates in the disc over the interval T_{rec} between outbursts, and accretes during an outburst of length ΔT, then by equating the masses we have $\dot{M}_{sec} T_{rec} = \dot{M}_{out} \Delta T$. Both T_{rec} and ΔT are readily measured from a lightcurve; in U Gem they are ~ 100 days and ~ 10 days respectively, leading to

$$\frac{\dot{M}_{out}}{\dot{M}_{sec}} \approx 10 \quad \text{and hence} \quad \frac{L_{out}}{L_{spot}} \approx 320.$$

Does this agree with the data? From the amplitude of the orbital hump in U Gem (Fig. 2.13) the bright spot emits roughly half the quiescent light, and so we expect $L_{out}/L_{quies} \approx 160$, which corresponds to a brightening by 5.5 magnitudes. This agrees with the observed outbursts (Fig. 5.1) to within the accuracy of the calculation, and so strongly supports the disc-instability hypothesis.

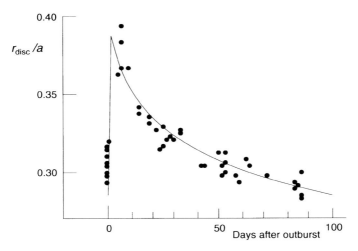

Fig. 5.4: Measurements of the disc radius in U Gem through the outburst cycle, obtained using the eclipse of the bright spot. The line illustrates the expected behaviour during a disc instability (see Fig. 5.13). Although the data are consistent with a disc instability, more extensive measurements, particularly in the early stages of outburst, would be beneficial. However, obtaining such data is not easy, due to the difficulty of catching dwarf novae on the outburst rise and the problems of measuring the disc edge when the bright spot is outshone by the disc. (Based on data by Joe Smak[4]).

in mass transfer, perhaps triggered by increased irradiation of the secondary during a disc instability, plays a supporting role in some dwarf-nova outbursts.

5.2 VISCOSITY IN AN ACCRETION DISC

Before discussing the theory of a disc instability I will first return to the viscosity in the disc. As discussed in Chapter 2, each annulus of the disc rotates at the local Keplerian velocity. Thus material which is further in orbits faster, and must slide past material further out (see Fig. 5.5). Any viscosity in the material opposes the sliding motion, and tries to force adjacent annuli to corotate. The outer annulus is sped up (giving it angular momentum) while the inner annulus is slowed down (draining its angular momentum); thus viscosity causes angular momentum to flow outwards. The majority of the material flows inwards, and releases gravitational energy, though a portion must spread outwards to carry the angular momentum (this is discussed mathematically in Box 5.2).

Although viscosity is essential to the operation of an accretion disc, the physical origin of the viscosity has been uncertain, defying theoretical investigation for many years. We know that molecules are sticky, attempting to form chemical bonds with their neighbours (this accounts for the viscosity of everyday materials such as treacle); however, discs are so diffuse that molecular viscosity is too feeble by a factor of a billion to explain their behaviour.

Box 5.2: Angular momentum exchange in an accretion disc

Lynden-Bell and Pringle[5] first described a simple two-particle model that illustrates the behaviour of an accretion disc. Consider two particles of mass m_1 and m_2 at radii r_1 and r_2 orbiting with Keplerian velocities around a large mass M. The angular momentum of the two (see Box 2.1) is

$$J = \sqrt{GM}\left(m_1 r_1^{1/2} + m_2 r_2^{1/2}\right)$$

which differentiates to

$$dJ = \frac{\sqrt{GM}}{2}\left(m_1 r_1^{-1/2} dr_1 + m_2 r_2^{-1/2} dr_2\right).$$

Conserving angular momentum ($dJ = 0$) thus requires

$$dr_2 = -\frac{m_1}{m_2}\left(\frac{r_2}{r_1}\right)^{1/2} dr_1.$$

Now the energy of the two-particle system is

$$E = -\frac{GM}{2}\left(\frac{m_1}{r_1} + \frac{m_2}{r_2}\right)$$

(the factor $1/2$ arises because half the lost gravitational potential energy is retrieved as orbital kinetic energy; see Box 2.1) and thus

$$dE = \frac{GM}{2}\left(\frac{m_1}{r_1^2} dr_1 + \frac{m_2}{r_2^2} dr_2\right).$$

Eliminating dr_2 then results in

$$dE = \frac{GM m_1 dr_1}{2r_1^2}\left[1 - \left(\frac{r_1}{r_2}\right)^{3/2}\right].$$

Thus the system sinks to a lower-energy configuration (while conserving angular momentum) given a negative dE. This requires a positive dr_1 if $r_1 > r_2$ and a negative dr_1 if $r_1 < r_2$. In other words, losing energy requires that the outer particle moves outwards and the inner particle moves inwards. The outward motion exceeds the inward motion (see the equation above relating dr_2 to dr_1) and so, in a many-body system, the overall effect is that a minority of the matter flows outwards, taking with it the majority of the angular momentum, while the majority of the material spreads inward.

Instead, theorists have proposed that discs are turbulent, so that adjacent annuli continually exchange blobs of material. This transfers angular momentum between the annuli and acts as a viscosity. In 1973 the theorists Shakura and Sunyaev[6] proposed that turbulent eddies must be smaller than the vertical height of the disc, H, and that, since gas molecules move at roughly the sound speed, eddies transfer material at up to the sound speed in the gas, c_s. Thus they parametrised the viscosity ν as

$$\nu = \alpha c_s H$$

where α, a number less than 1, denotes the size of the viscosity as a fraction of the limiting case. By using this 'alpha viscosity' theorists had an equation which could be combined with equations for gas dynamics to build model accretion discs.

Such discs are commonly called 'alpha discs' and are the standard theoretical model. They are 'thin discs', in that the height of the disc is very much smaller than the radius. They are slightly concave, flaring at their outer edges,* and their total mass is negligible in comparison with that of the central white dwarf. The speed of the orbital motion (of order 1000 km s^{-1}) vastly exceeds the sound speed in the gas (~ 10 km s^{-1}), but the speed at which material drifts inward (~ 0.3 km s^{-1}) is much slower again.

By comparing alpha-disc models to lightcurves of dwarf novae, one can deduce that α has to be about 0.01–0.05 in the cold disc of a quiescent dwarf nova, jumping to about 0.1–0.5 during an outburst, but the alpha prescription itself gives no clue to the cause of the turbulence required to create the viscosity. Indeed it can be shown that, considering only hydrodynamics, Keplerian discs are stable and tend to damp out turbulence, even if turbulence is artificially added to a model.

5.2.1 Magnetic turbulence

During the 1990s a theory of turbulence driven by magnetic instabilities was developed by Steven Balbus and John Hawley,[7] and is now widely regarded as having solved the viscosity problem (although there are competing ideas and the matter is not finally settled). To understand it, some properties of the interaction of a magnetic field with an ionised gas need to be considered. Given the intimate connection between electricity and magnetism, the free electric charges in ionised material ensure a strong coupling between the material and any magnetic fields. While charged particles readily flow along magnetic field lines, they cannot easily cross field lines. Furthermore, field lines are dragged along by the motion of ionised matter. Thus, to a large extent, the field lines and material are frozen together and move in harmony.

Now consider a field line connecting two bubbles of ionised gas, one at a slightly larger radius than the other. The inner bubble orbits faster, so the field line becomes stretched. It tries to pull the bubbles back together, which speeds up the outer one, giving it angular momentum, and slows down the inner one, removing angular momentum. But the gain of angular momentum causes the outer bubble to move outwards, and the loss of angular momentum causes the inner bubble to fall inward,

---

*$H \propto R^{9/8}$, though this doesn't necessarily apply if α varies with radius.

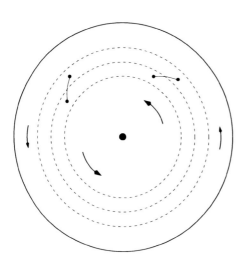

Fig. 5.5: A schematic of viscosity in an accretion disc. Blobs of material in neighbouring annuli are connected by magnetic fields, but the inner annulus moves faster than the outer annulus. As explained in the text, this stretches and amplifies the magnetic field, leading to magnetic turbulence.

so the net effect is that the field line is stretched even more. The stretched field line is equivalent to a stronger magnetic field, so small initial fields are greatly amplified by this effect. Ultimately, as the field grows, orderly flow breaks up into magnetic turbulence, a process now referred to as the Balbus–Hawley instability.

To appreciate this, consider a vertical field line (one which is perpendicular to the disc plane) possessing small deviations to larger and smaller radii (Fig. 5.6). As just explained, the outward deviations move further outward and the inward deviations move further inward, amplifying the deviations, until loops of field line penetrate to larger and smaller radii. But material largely follows the field lines, and so the magnetic loops efficiently transport material to different radii. The resulting turbulence is exactly what Shakura and Sunyaev envisaged as the driving mechanism of an accretion disc.

5.3 THE THERMAL INSTABILITY

The Balbus–Hawley instability requires a strong coupling between the magnetic field and the disc material. This occurs when the disc is hot and ionised, containing many free electric charges. But in a cold disc, electrons combine with nuclei to create electrically neutral atoms, and the material cannot couple to the magnetic field, shutting down the Balbus–Hawley instability. This is the essential difference between the cold, low-viscosity disc of a quiescent dwarf nova and the hot, high-viscosity disc of outburst.

A mechanism that can flip the disc between hot and cold states can thus cause a dwarf-nova outburst. Ironically, this mechanism again arises from the transition from neutral atoms to ionised plasma. Photons are, of course, packets of electromagnetic radiation and so interact strongly with charged particles and only weakly with neutral atoms. Cold, un-ionised material (for instance, hydrogen below a few thousand Kelvin) thus has a low opacity to radiation (where 'opacity' means the

Sec 5.3   The thermal instability   63

Larger r   Smaller r
←   →

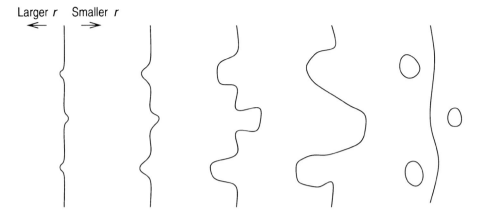

Fig. 5.6: The sequence from left to right illustrates the growth of the Balbus–Hawley instability. The deviations in a vertical field line are amplified, increasing the strength of the field until it 'reconnects' to dissipate energy. Since the ionised material follows the field lines, bubbles of gas are transported to different radii.

ability to obstruct the flow of radiation). When hydrogen is heated to ∼ 7000 K, however, some electrons break loose from their atoms and the gas becomes partially ionised, boosting the opacity (in particular, some of the free electrons combine with hydrogen atoms to create H^- ions, which are particularly efficient at absorbing photons). If one tries to heat the gas further, most of the energy goes into creating more ions rather than heating the gas, so that a small increase in temperature produces a large increase in ionisation. The overall effect is that, in a partially ionised hydrogen gas, opacity is extremely sensitive to temperature, increasing as its tenth power! Between 1979 and 1981, through pioneering work led by Reiun Hoshi, Jim Pringle and Friedrich Meyer with Emmi Meyer-Hofmeister,[8] it was realised that this could cause discs to be unstable.

To appreciate the relevance of the opacity we need to discuss the stability of an accretion disc. Suppose that the temperature in one part of a disc fluctuates upwards a little. This would increase the random motion of the gas particles and so increase the viscosity; the increased viscosity increases the flow of material inwards, emptying the region (this is not fully replenished from larger radii since they still have the lower viscosity). The lower density reduces the viscous heating, and so the temperature drops back to normal and the region fills up again. Thus the situation is stable.

Now reconsider the fluctuation when hydrogen is partially ionised. The rise in temperature causes a huge increase in opacity; the opacity traps the heat created by viscous interaction, and so the temperature climbs further. Yes, the viscosity is climbing, emptying the region and so reducing the amount of heat produced by viscous interactions, but this reduction is far outweighed by the trapping of energy within the disc. Thus the temperature continues to climb until the hydrogen is completely ionised. At this point the opacity loses its extreme sensitivity to temperature, and the disc settles into a new equilibrium state at a much higher temperature.

In the hot, highly viscous state, the inward flow of material through the disc exceeds that entering the disc from the mass-transfer stream. Thus the new hot state cannot be sustained for long, and the disc must eventually return to its original condition.

The changes outlined above can be considered as a cycle on a plot of the disc's surface density versus its surface temperature (Fig. 5.7). In quiescence the dwarf nova sits near point A, at which the disc is stable. However, suppose that the mass transfer from the secondary is greater than the flow through the low-viscosity disc. The disc will fill up, gradually increasing its surface density. The greater viscous heating of the more-massive disc raises its temperature (see Box 5.3), and the disc moves toward B on the temperature–density plane. At B, where ionisation sets in, any further rise in surface density produces a runaway rise in temperature. Since the timescale for heating is far shorter than the time for viscous exchange of material, the surface density stays roughly constant as the temperature increases until hydrogen is completely ionised (C in Fig. 5.7). A new equilibrium is established in which the increased luminosity resulting from the higher temperature is sustained by a much higher inward flow of material driven by the higher viscosity of the ionised state.

However, this flow is greater than is resupplied by the secondary. The surface density therefore drops, causing a corresponding drop in temperature until the disc reaches D. At this point hydrogen again becomes partially ionised, the high opacity returns, and the heat to maintain the high surface temperature can no longer escape.

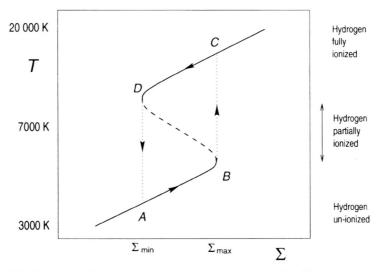

Fig. 5.7: The cycle of a dwarf nova plotted as surface density (Σ = mass per area of disc) versus disc temperature. The S-curve of the T–Σ relation forces the disc to follow the limit cycle from A → B → C → D → A, as discussed in the text.

The temperature plummets, the ions recombine, the viscosity drops, and the disc returns to the cold, quiescent state.

5.3.1 Heating and cooling waves

I have so far discussed the outburst as if the whole discs acts in unison. In fact, an instability develops first at one annulus in the disc. The higher viscosity then spreads hot material from that annulus into adjacent annuli, pushing them over the instability threshold, and so on in a domino effect. The resulting 'heating wave' spreads through the disc, sending it all into outburst.

The evolution of the outburst depends on the distribution of material in the disc when the outburst starts, and on the radius at which the outburst is triggered. The following (rather technical) discussion is aided by the plots (Fig. 5.8) of surface density (Σ) as a function of radius. The first point is that the two critical surface densities — $\Sigma_{\max}$, the maximum density on the lower branch of the T–Σ relation (Fig. 5.7), and $\Sigma_{\min}$, the minimum density on the upper branch — both depend on the radius in the disc. The reason for this is that, because the gravitational potential well steepens at smaller radii, the flow of material liberates more energy and thus causes a higher temperature when it is further in (the $T \propto r^{-3/4}$ law of

---

Box 5.3: The origin of the S-curve

The flow of matter through a disc is the product of the viscosity, ν, and the surface density (mass per area), Σ. The mass flow determines the energy radiated away and thus sets the surface temperature (see Box 3.1), so that

$$\dot{M} \propto T_{\mathrm{surf}}^4 \propto \nu \Sigma.$$

In the alpha prescription, $\nu = \alpha c_{\mathrm{s}} H$, and since the sound speed $c_{\mathrm{s}} \propto \sqrt{T}$ and also $H \propto \sqrt{T}$ we can write

$$T_{\mathrm{surf}}^4 \propto \alpha T_{\mathrm{mid}} \Sigma$$

where we have distinguished between the surface temperature and the temperature in the mid-plane, where most of the viscous dissipation is occurring.

If the opacity, κ, is not a strong function of temperature then outward heat transport is steady, and $T_{\mathrm{mid}} \propto T_{\mathrm{surf}}$, which leaves $\alpha \Sigma \propto T_{\mathrm{surf}}^3$. Thus an increase in surface density produces an increase in temperature (treating α as constant).

However, in the partially ionised regime, where $\kappa \propto T^{10}$, the opacity traps the radiation and decouples the mid-plane temperature from the surface temperature so that $T_{\mathrm{mid}} \propto T_{\mathrm{surf}}^{10}$. Hence

$$\alpha \Sigma \propto T_{\mathrm{surf}}^{-6}.$$

Thus an increase in temperature equates to a *reduction* in surface density (the slope of the T_{surf}–Σ plot is negative), even accounting for reasonable changes of α with temperature.[8] The overall result is the S-curve illustrated in Fig. 5.7.

66 Discs and outbursts

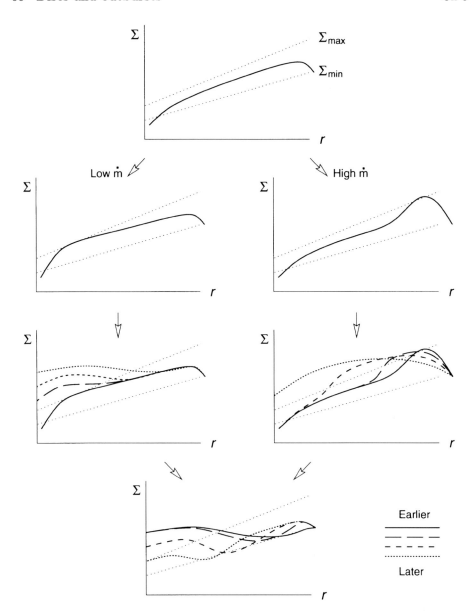

Fig. 5.8: The disc's surface density (Σ) versus its radius over the outburst cycle (see the text for fuller details). The disc starts in the state of the upper panel. A low mass-transfer rate then triggers an inside-out outburst while a high mass-transfer rate triggers an outside-in outburst (the evolution is shown by the sequence of lines in different styles). Near the end of the outburst the disc state is similar for both outburst types, leading to a cooling wave that moves inward from the outer disc. The schematic plots are based on computations by Shin Mineshige and Yoji Osaki.[9]

Box 3.1). Thus, since the flow of matter depends on the density (flow $\propto \nu\Sigma$), the disc can support a higher Σ at larger radii without hitting the stability limit.

The effect of the previous outburst is to leave the disc in a state where Σ lies between the two critical values (Fig. 5.8). It then fills up during quiescence, in a manner dependent on the rate of mass transfer from the secondary. At low transfer rates the material has time to diffuse inward viscously, and so accumulates at small radii. Σ then hits Σ_{max} first in the inner disc. At high transfer rates the material does not have time to diffuse inwards, piles up at large radii, and so triggers the instability in the outer disc (second panels of Fig. 5.8).

When the trigger is in the inner disc, the heating wave pushes outward, creating an 'inside-out' outburst. As each new annulus switches to high viscosity, the matter flowing into the inner regions increases, boosting their surface density and increasing the rate of accretion onto the white dwarf (left-hand panels of Fig. 5.8).

In the opposite case of an 'outside-in' outburst (right-hand panels of Fig. 5.8), the heating wave travels inward, bringing with it a density enhancement. The final result, in the 'steady state' of outburst, is similar in the two cases: the density profile is reversed and Σ is now greater in the inner disc.

The enhanced accretion rate of outburst drains the disc of its material until Σ is reduced to Σ_{min} at some annulus, which is always in the outer disc since Σ_{min} in highest here. This annulus drops out of outburst and a 'cooling wave' spreads inward. The regions just inside the cooling wave have a higher viscosity than the regions just outside it, and so push material outwards across the cooling front. This reduces Σ in the just-inside annulus to Σ_{min}, so that it too drops out of outburst and the cooling wave sweeps inward (lowest panel of Fig. 5.8), sucking material back to larger radii as it passes. Eventually the whole disc is returned to quiescence, leaving

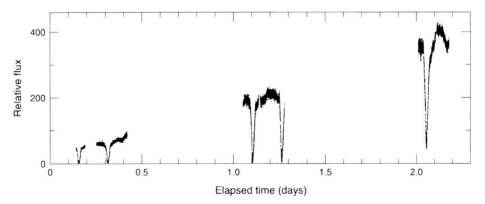

Fig. 5.9: A lightcurve of IP Peg on the rise to outburst. Initially the eclipses are near total, implying that the bright regions are small enough to be entirely eclipsed by the secondary. Thus the outburst must be confined to the inner disc. Later, the flux remaining at mid-eclipse implies that the outburst has spread to the outer disc, which can never be entirely eclipsed (see Fig. 2.10). This was therefore an 'inside-out' outburst. (Data courtesy of Natalie Webb and Tim Naylor.[10])

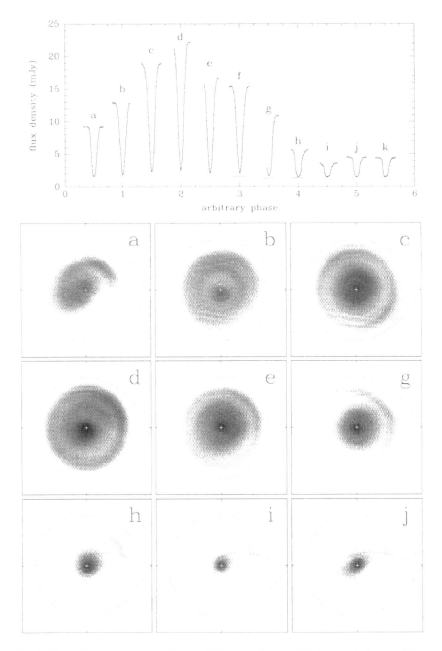

Fig. 5.10: Eclipses over an outburst of the dwarf nova EX Dra and the resulting disc maps obtained by eclipse-mapping (see Box 3.2). On the outburst rise (a → d) the disc expands. On the decline (d → g) the cooling wave moves inward so that the inner regions remain bright longer. Back in quiescence (h → j), the flux due to the mass-transfer stream and bright spot again becomes significant. (Figures by Raymundo Baptista and Sandi Catalán.[11])

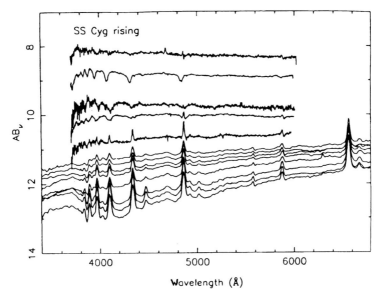

Fig. 5.11: Spectral changes as SS Cyg rises from quiescence (lowest spectrum) to outburst (uppermost spectrum). The continuum becomes bluer as the disc heats, while the emission lines of quiescence evolve into shallow absorption features. One interpretation is that the disc is optically thin in quiescence, turning optically thick in outburst due to the much greater ionisation of the material. However, there is evidence that discs are also optically thick in quiescence; for instance, the orbital hump, seen only for half the orbit (e.g. Fig. 2.13), implies that the bright spot cannot be seen through the disc. Thus the emission lines of quiescence might arise from an optically thin corona surrounding the disc, which is outshone by the optically thick disc itself during outburst. (Figure by Keith Horne.[12])

a density profile similar to that with which it started.

The enhanced accretion onto the white dwarf lasts only from the time when the heating wave reaches the innermost disc to the time when the cooling wave does likewise. In this period only $\sim 10\%$ of the matter in the disc has time to accrete. Much of the remaining material simply flows inward (due to the heating wave) and then outward again (due to the cooling wave).

5.3.2 Outburst shapes

Outburst lightcurves, even in the same star, are not uniform. Fig. 5.12 shows a year-long segment of SS Cyg's lightcurve. Notice that some outbursts are shorter than others, and that while most have very rapid rises and slower declines, at least one has a slow rise and thus a more symmetric shape. The reasons for such differences must involve the radius at which the outburst is triggered and the distribution of material in the disc left by the previous outburst. Our current understanding of how these relate to outburst shapes is as follows.

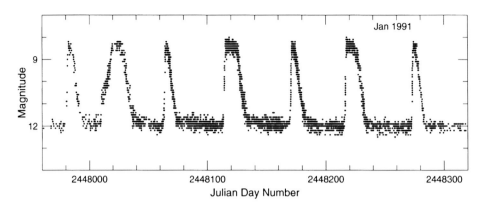

Fig. 5.12: A 1-yr section of SS Cyg's lightcurve showing a range of outburst shapes. The outburst rises are either rapid (~ 2 days) or slow (~ 8 days); there might or might not be a plateau phase lasting ~ 10 days, but the declines all take ~ 8 days. (Data compilation by the AAVSO.)

- The rapid rises are caused by outside-in outbursts. The inward-running heating wave travels rapidly because (1) viscosity causes more material to flow inwards than outwards (Box 5.2); (2) outer radii contain a higher surface density (Fig. 5.8), so spreading this material has a large effect on annuli further in, and (3) annuli at smaller radii are smaller, so that funnelling material inwards boosts the density. The three combined allow an inward-running heating wave to overwhelm the next annulus quickly and easily, so that the wave runs rapidly through the disc. Accretion increases suddenly when the heating wave and its accompanying flood of material reaches the white dwarf.

The speed of the heating wave is $\alpha_{\text{hot}} c_s$ (where α_{hot} means α on the hot side of

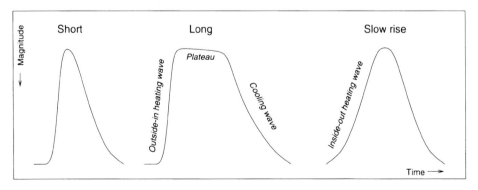

Fig. 5.13: Schematic outburst types, for comparison with the data in Fig. 5.12. The different shapes are explained in the text. Since all observed outbursts are subtly different, one could, should one wish, divide them into many more types.

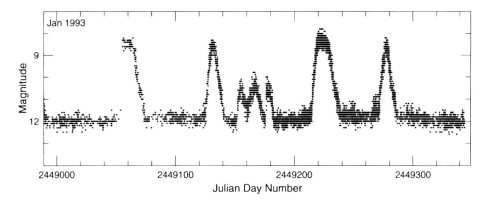

Fig. 5.14: Occasionally SS Cyg shows anomalous outbursts, periods of small-amplitude, chaotic brightness variations. Presumably, the different sections of disc are acting independently, instead of in concert, with cooling and heating waves moving through different regions of the disc simultaneously. Modelling codes do not yet have the sophistication to reproduce this behaviour. (Data compilation by the AAVSO.)

the instability), which combined with knowledge of the rise time and the distance travelled (the disc radius) allows us to estimate $\alpha_{\rm hot}$ as ~ 0.1–0.3.

• The slow rises are caused by inside-out outbursts. The three factors above hamper an outward-running heating wave, rather than assisting it, so that it travels more slowly. Accretion rises gradually as more and more of the disc participates in sending extra material inward.

• The short outbursts occur when the heating wave fails to propagate right to the disc's outer edge (even if the trigger occurs at large radii, a fall-off in density towards the edge might prevent the outermost regions being pushed over $\Sigma_{\rm max}$). The remaining cold region then sucks material out of the hot region, initiating a cooling wave almost immediately, and shutting down the outburst.

• Plateaus occur when the entire disc enters outburst, and sustains it, with matter draining onto the white dwarf, until Σ drops to $\Sigma_{\rm min}$ in the outer disc (the slight decline during this phase corresponds to the decline from C to D in Fig. 5.7). Some authors argue that enhanced mass transfer in outburst due to irradiation of the secondary helps to sustain plateaus.

• Long and short outbursts have a tendency to alternate. When the outermost regions don't participate in a short outburst, they are left with more material, which increases the likelihood that they will participate in the next instability.

• Since the cooling wave always originates in the outer disc and moves inwards, the declines of all outbursts are similar.

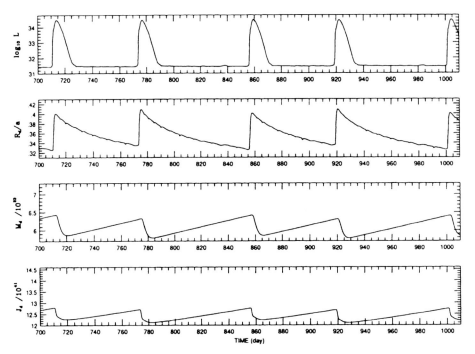

Fig. 5.15: Model simulations of outbursts, computed by Susumu Ichikawa and Yoji Osaki.[13] The panels show (from the top) the luminosity, the disc radius, the mass of the disc, and its angular momentum. (Details of such models are beyond the level of this book, but for an introduction to the equations involved consult *Accretion Power in Astrophysics* by Frank, King and Raine.)

5.4 NOVALIKE VARIABLES AND Z CAM STARS

The thermal instability can operate only if the mass-transfer rate into the disc corresponds to a temperature in the partially ionised zone (between B and D of Fig. 5.7; recall also that higher mass flow produces a higher temperature; Box 3.1). Since no stable state exists at this temperature, the disc fluctuates between higher and lower mass-flow states, effectively averaging itself to the flow from the secondary.

If the flow were to be so low that the steady-state temperature would be insufficient to ionise hydrogen (corresponding to the lower branch Fig. 5.7), the disc would never undergo an outburst and would simply remain quiescent. We don't know of any such star: either gravitational radiation drives mass transfer sufficiently that this never happens, or such stars are so faint and unremarkable (they don't draw attention to themselves by outbursts) that we have overlooked them.

Likewise, if the mass transfer is so high that it corresponds to the hot, viscous upper branch of Fig. 5.7, the disc will be stuck permanently in outburst (this limit is about 10^{14} kg s^{-1}). We call such stars *novalike variables*. This nomenclature needs an explanation. Early observers knew about novae (see Chapter 11) and dwarf

Sec 5.4 Novalike variables and Z Cam stars 73

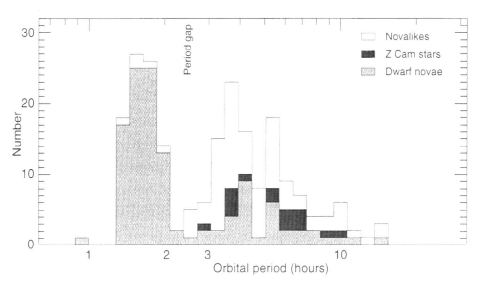

Fig. 5.16: The orbital periods of dwarf novae, Z Cam stars and novalike variables, from a compilation by Hans Ritter and Uli Kolb.[14]

novae, and also found stars which were similar to the remnants of past novae. They called these 'nova-like variables', abbreviated to 'novalikes', supposing (correctly) that novalikes and old novae were the same type of star, the only distinction being whether one had been observed to undergo a nova eruption. Nowadays, the term 'novalike' is applied to any cataclysmic variable with a mass-transfer rate sufficient to sustain the disc on the hot side of the instability (although some authors retain the old-nova/novalike distinction). Novalikes are also sometimes referred to as 'UX UMa' stars, after an exemplar system.

Fig. 5.16 shows the distribution of the orbital periods of the known novalikes compared to those of dwarf novae. The novalikes are nearly all found above the period gap. This accords with the theory that high mass-transfer rates are driven by magnetic braking, and that this mechanism does not operate below the gap (Chapter 4). Nearly all systems below the gap are dwarf novae, presumably undergoing mass transfer at the lower rate appropriate to gravitational radiation. Note, though, that both dwarf novae and novalikes occur at all orbital periods, which implies that mass-transfer rates can vary markedly between different systems with the same orbital period. We return to this fact in Chapter 12.

Z Cam stars are poised on the borderline between dwarf novae and novalikes. Their lightcurves (Fig. 5.17) show periods of rapid outburst activity interspersed with periods of novalike steadiness, referred to as *standstills*. The supposition is that a small change in mass-transfer rate can switch them from one behaviour to the other.

Notice that standstills are always initiated by an outburst. Something — perhaps heating of the secondary by the enhanced irradiation of outburst — then boosts the mass-transfer rate. The disc reaches a new equilibrium (somewhere along the $C \to D$ branch of Fig. 5.7) with enhanced irradiation sustaining enhanced mass transfer. The Z Cam star acts as a novalike for a while, until something — perhaps a star-spot passing across the L_1 point — causes the mass-transfer rate to drop again, breaking the feedback loop. The star then drops out of outburst (notice that standstills always end with a decline to quiescence; Fig. 5.17). But the mass transfer is still very high for a dwarf nova; the disc fills up quickly, and outbursts recur in quick succession separated by short quiescent intervals. Eventually, one of the outbursts triggers another standstill.

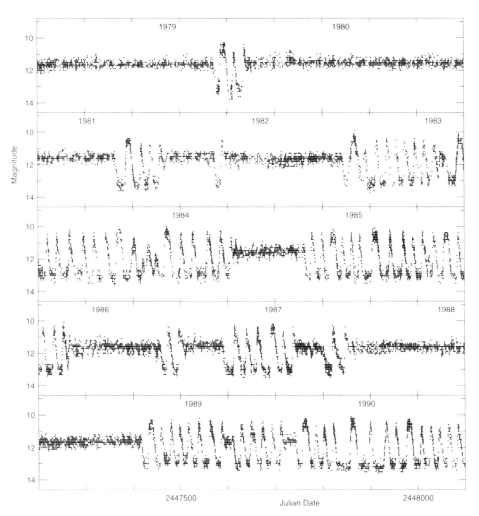

Fig. 5.17: Twelve years in the life of Z Cam, compiled by the AAVSO. Tickmarks are at 100-day intervals.

Chapter 6

Elliptical discs and superoutbursts

The outbursts of stars such as U Gem and SS Cyg, discussed in the last chapter, show a range of different lengths and shapes. However, it is always possible to find outbursts which are intermediate between types, implying that the different shapes are merely variations on a theme.

One class of dwarf nova, though, shows outbursts that clearly fall into two categories. For example, consider the lightcurve of VW Hyi (Fig. 6.1). While most outbursts last 3 days, and are similar to those in U Gem and SS Cyg, there are also less frequent *superoutbursts*, lasting ~ 14 days. The superoutbursts are regular (more so than the normal outbursts) and are slightly brighter. Often it appears that superoutbursts are triggered by normal outbursts, revealed by a slight hiatus before the superoutburst brightens to maximum.

Photometry of superoutbursts reveals another peculiarity: a hump-shaped modulation that appears near superoutburst maximum. The humps do not recur with the orbital cycle, but with a period a few percent longer than the orbital cycle (see Fig. 6.2). These *superhumps* continue until the star returns to quiescence, although their period usually drifts to slightly shorter periods over this time.

Stars which show both superoutbursts and superhumps are referred to as SU UMa stars (we know of no system showing only one of these characteristics). For distinction, dwarf novae which are neither SU UMa stars nor Z Cam stars are called U Gem stars, or sometimes SS Cyg stars. (This habit of naming subclasses of cataclysmic variable after an exemplar of the type can confuse those new to the field; the advantage of such labels is that they are based on observational characteristics, allowing us to change our minds about the physical relations between subclasses without having to change the nomenclature. The reader should also be prepared for slight differences in the subclass definitions between different authors.)

6.1 ELLIPTICAL DISCS

Nicholas Vogt first proposed that superhumps were caused by the disc becoming elliptical during superoutburst.[1] He suggested that such a disc would precess, meaning that the direction in which the disc was elongated would gradually rotate, on a

76 Elliptical discs and superoutbursts

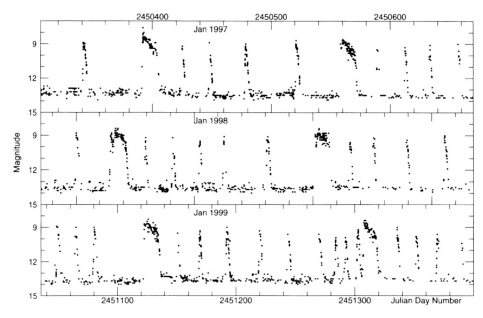

Fig. 6.1: A 3-yr section of VW Hyi's lightcurve showing both normal outbursts and superoutbursts. (Data by the Royal Astronomical Society of New Zealand.)

timescale much longer than the orbit (in the same way, the axis of a spinning top precesses, but more slowly than it spins). The long, precessional period of the disc would then interact with the orbital cycle to create a new periodicity — the superhump. To see this, consider the secondary lining up with the elongated radius of an elliptical disc, which precesses with a period of a few days. One orbital cycle later, the secondary would again be in the same place, but the elongation would have

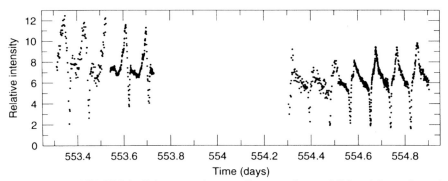

Fig. 6.2: DV UMa's lightcurve shows eclipses and an additional 'superhump' modulation. Since the superhump maxima occur at different phases with respect to the eclipse on different nights, the superhump period is slightly longer than the orbital period. [Data by the Center for Backyard Astrophysics[2] (the quality varies because data from different telescopes were combined).]

moved on slightly, so the secondary would have to continue a little further before they again lined up. Any interaction between the two (such as the mass-transfer stream interacting with the disc) would thus occur at a slightly longer 'superhump' period that can be considered the *beat period* between the orbital and precessional periods (the relation is discussed mathematically in Box 6.1).

Although elliptical discs were widely suspected of involvement in superoutbursts, the reason for the disc becoming elliptical and the cause of the extra light of the superhump remained mysterious until 1988. In that year a graduate student at Oxford, Rob Whitehurst, published a computer simulation in which the disc was modelled as a mass of interacting particles.[3] He found that, provided the mass ratio obeyed $q \lesssim 0.33$, the disc would become elliptical and begin to precess. The origin of the ellipticity lay in the disc's interaction with the secondary star. To understand this we need to discuss the tidal torque exerted by the secondary star in more detail.

6.2 TIDAL TORQUES AND RESONANCES

In the inner disc, orbits about the white dwarf are circular. However, towards the edge of the disc the orbits are increasingly distorted by the presence of the secondary and become non-circular. Material is pulled towards the secondary, causing a slight bulge. The disc orbits faster than the secondary, so the bulge tries to move ahead of the secondary (viscous interactions, caused by collisions between particles on adjacent, non-circular orbits, allow the disc to behave as a bulk fluid) but the

---

Box 6.1: 'Beating' of two periods

Consider an elliptical disc precessing slowly, with a period P_{prec}, in the same direction as the orbital motion ('prograde'). The number of orbital cycles occurring within a time P_{prec} is thus $P_{\text{prec}}/P_{\text{orb}}$. The number of interaction or 'superhump' cycles in this time will be one fewer (to visualise this, consider the orbital cycle as winding up a ball of string that is attached to the disc; every precessional cycle turns the ball once and unwinds one loop) and so

$$P_{\text{prec}}/P_{\text{sh}} = P_{\text{prec}}/P_{\text{orb}} - 1$$

or

$$\frac{1}{P_{\text{sh}}} = \frac{1}{P_{\text{orb}}} - \frac{1}{P_{\text{prec}}}$$

where P_{sh} is the superhump period. This relation applies generally to a third period generated by the difference of two other periods, and the derived period is referred to as the 'beat period'. We will meet this concept again when discussing magnetic systems in Chapter 8. (Purists might insist that the derived period should only be called the beat period if it is much longer than the first two periods. However, since the mathematical relation is identical in any case, and since it is a useful term for the interaction period, I will adopt this usage despite possible howls of outrage.)

---

78 Elliptical discs and superoutbursts

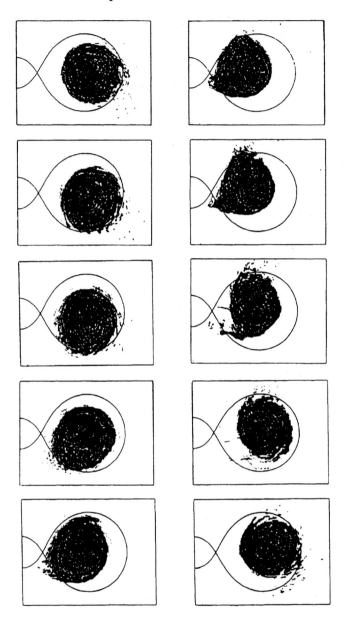

Fig. 6.3: A model of an eccentric precessing accretion disc, computed from the hydrodynamics of interacting particles (by Masahito Hirose and Yoji Osaki[4]). The snapshots show the disc in steps of 0.1 in phase round the superhump cycle (the sequence is top left to bottom left then top right to bottom right). The secondary's Roche lobe is seen on the left. As seen from the inertial frame, the eccentricity precesses in the same direction as the orbital motion ('prograde') but with a longer period of a few days; as seen from the faster-moving secondary, the eccentricity would appear to rotate in the opposite sense ('retrograde').

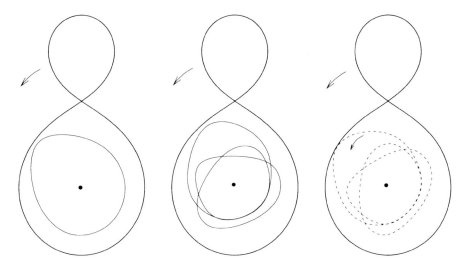

Fig. 6.4: In systems with high mass ratios the orbits in the outer disc are distorted by the presence of the secondary, and form a tidal bulge slightly ahead of the secondary (left panel). If a resonance sets in, the orbit is no longer fully periodic; a particle orbits three times in the time the secondary orbits once (middle panel) producing a three-petalled orbit. Such orbits precess slowly (right panel), so the eccentricity of the disc precesses.

gravitational pull of the secondary holds it back. Thus, in equilibrium, the bulge lies slightly ahead of the line joining the white dwarf to the secondary. (For the same reason, the tidal bulge of the Earth's oceans points towards, but slightly ahead of, the Moon.) The pull of the secondary slows the material in the bulge, reducing its angular momentum, and so providing the drain of angular momentum necessary for the inward flow of material through the disc. This 'tidal torque' stops the disc growing and sets the outer disc radius.

The above situation, though, is unstable if resonances come into play. A periodic motion has a natural period of oscillation, and a resonance occurs if a force driving the motion has the same period (a familiar example is a child's swing, where pushing in time with the natural frequency of the swing causes the amplitude to increase).

The motion of a particle in a non-circular orbit can be thought of as an 'in-out' radial motion superimposed on a circular orbit. If the gravitational 'kick' of the secondary star resonates with the radial motion, it will enhance the radial component of the motion, driving the outer parts of the disc elliptical.

Since the Keplerian orbit at the disc edge is much faster than the secondary's orbit, the material cannot directly resonate with the secondary. However, suppose that disc material orbits twice per (secondary) orbital period; it would receive a kick every second orbit, which could still drive the disc elliptical (an analogy would be pushing a swing every second pass). The radius of such orbits, though, would be bigger than the white dwarf's Roche lobe in most cataclysmic variables, whereas for a disc to achieve resonance, these orbits have to be inside the tidal truncation radius. This is only possible for extreme mass ratios of $q \lesssim 0.025$, which are attained only by systems that have evolved past the period minimum (see Chapter 4).

However, the next resonance, in which the disc material orbits three times for every orbital cycle, is attainable for systems with $q \lesssim 0.3$. (Remember that by Kepler's law a smaller orbit has a shorter period; furthermore, the more extreme the mass ratio, the more room there is for the disc to grow, and the interplay between the two results in the limit on q, as discussed in Box 6.2). This resonance is thought to be the origin of elliptical discs in superoutburst (higher-order resonances occur at still smaller radii, but the effect becomes weaker and weaker, as is obvious from our analogy with the swing).

6.2.1 The effect of the orbital period

The limit on mass ratio explains the difference between the SU UMa and U Gem subclasses of dwarf nova: one has mass ratios smaller than the limit and the other has larger mass ratios. Since the mass of the secondary star increases as the orbital

---

Box 6.2: Mass ratios for resonance

The radii at which the orbits in the disc resonate with the secondary can be found as follows. Kepler's law (Box 2.1) can be rearranged to give the separation a as

$$a^3 = P_{\text{orb}}^2 M_1 (1+q) G / 4\pi^2.$$

Similarly, ignoring the perturbation of the secondary, an orbit in the disc with period P has a radius

$$r^3 = P^2 M_1 G / 4\pi^2$$

and thus

$$\frac{r}{a} = (P/P_{\text{orb}})^{2/3} (1+q)^{-1/3}.$$

The maximum disc radius is set by the tidal limit — the point at which orbits start intersecting — which is approximated by (Box 2.4)

$$\frac{r_{\text{tidal}}}{a} = \frac{0.6}{1+q}.$$

Thus for the 2:1 resonance in which $P = \frac{1}{2} P_{\text{orb}}$ there is no value of q for which r is inside r_{tidal} (in fact, the tidal limit formula becomes inaccurate for $q < 0.03$ and it is possible that this resonance occurs for $q \lesssim 0.025$).

However, for the 3:1 resonance, $P = \frac{1}{3} P_{\text{orb}}$, r is inside the tidal limit for $q \lesssim 0.3$. This treatment ignores that fact that the secondary motion should resonate with a *precessing* disc orbit, not the simple Keplerian orbit; including this effect[5] the resonance zone is at $r \sim 0.46a$.

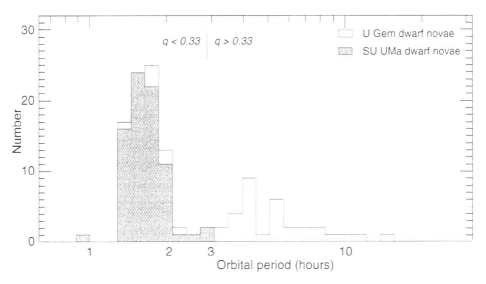

Fig. 6.5: The orbital periods of U Gem and SU UMa subtypes of dwarf nova (from a compilation by Hans Ritter and Uli Kolb[6]). The line marks the orbital period corresponding to a mass ratio $q = 0.33$, assuming that the white-dwarf mass is 0.7 $M_\odot$ and that the secondary obeys the period–mass relation from Appendix A. Some of the systems which are currently classified as U Gem types at short orbital periods are poorly studied; further observations of them may detect superhumps, leading to their reclassification as SU UMa stars.

period increases (the longer the period the larger the Roche lobes and thus the bigger the radius of the secondary), stars with sufficiently small mass ratios occur preferentially at short orbital periods. In fact the divide of $q \approx 0.3$ occurs near the period gap. Thus dwarf novae below the gap are almost all of SU UMa type, whereas those above the gap are nearly all U Gem type (the division by orbital period is of course not rigid since q also depends on the white-dwarf mass). This is illustrated in Fig. 6.5.

Confirmation of the above theory results from considering the period of the superhump in binaries at different orbital periods. The rate at which the resonant orbits precess depends on the mass ratio,* and since the superhump period is set by the beating of the precession and the orbit, the superhump period-excess increases with mass ratio. Since the mass ratio is often difficult to measure, whereas the orbital period is easy, one can use the fact that mass ratio increases with orbital period, and instead plot superhump period-excess against orbital period, as in Fig. 6.6.

---

*Patterson[7] quotes $P_{\rm orb}/P_{\rm prec} \approx 0.233 q (1+q)^{-1/2}$, and this is used in Fig. 6.6.

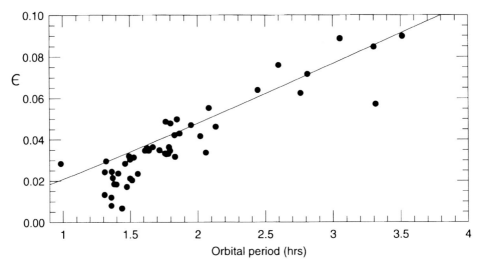

Fig. 6.6: The superhump period excess, ϵ, defined as $\epsilon = (P_{sh} - P_{orb})/P_{orb}$, versus orbital period. The line uses an estimate for the variation of ϵ with q (footnote on page 81), together with the assumption that the secondary masses depend on P_{orb} as given in Appendix A, and that the white dwarf masses are all 0.7 $M_\odot$. Since all of these inputs are uncertain, the line is a surprisingly good fit! It also ignores the fact that the disc is a fluid, and not a collection of free particles.[7]

6.2.2 The superhump lightsource

The gravitational kick of the secondary, resonating with the orbits in the disc, drives the disc elliptical, but why does this produce the extra light we see as a superhump? The answer is that the elliptical orbits are no longer parallel to those of their neighbours. The tidal stresses cause the orbits to intersect, and the resulting collisions dissipate energy; as the secondary sweeps past a tidal bulge, the disc brightens where the orbits intersect.

Observational confirmation of this was provided by Darragh O'Donoghue.[8] He observed eclipses of the SU UMa star Z Cha, first when the eclipses coincided with the maximum light of the superhump, and also when the eclipses coincided with the minimum of the superhump. By subtracting the two he was left with an eclipse profile of the superhump light. Using the techniques of eclipse mapping (Box 3.2) he showed that the superhump light came from the rim of the disc, in three distinct regions (see Fig. 6.7). The three regions are where, according to calculations, orbits intersect and dissipate energy.

6.3 THE EVOLUTION OF SUPERHUMPS

Superhumps usually appear round the time of the peak of the outburst, when they often display their maximum amplitude. As the outburst declines, the hump amplitude decreases and the pulse profile changes (see Figs. 6.8 and 6.9). Furthermore,

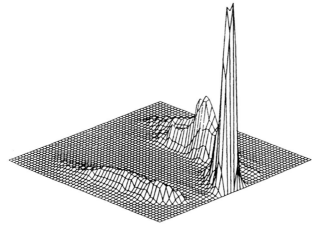

Fig. 6.7: The distribution in the disc of the extra light of the superhump. The white dwarf is in the centre of the grid, the red dwarf is off the grid to the lower-right, and the superhump light comes from three regions along the rim of the disc, where particle orbits intersect. This plot, by Darragh O'Donoghue, is derived from eclipse maps of Z Cha during superoutburst.[8]

the superhump period drifts, decreasing by $\approx 0.6\%$ per magnitude of decline.[9] The explanation for this is probably the gradual emptying of the disc, which causes its radius to shrink, and so lengthens the precession period.

Late on the decline the superhumps change abruptly, becoming roughly anti-phased with the earlier superhumps, so that hump maximum occurs where a minimum would be expected. This behaviour is best displayed by comparing the observed times of maxima with the predicted times, calculated by extrapolating earlier superhumps and assuming no change. The result is an 'observed' minus 'calculated' or '$O-C$' diagram, such as that in Fig. 6.10 (see also Box 6.3).

We currently don't understand the change to *late superhumps*, as they are called. One can speculate that the draining of material from the disc forces a redistribution that results in the eccentricity changing phase. Another possibility is that late superhumps result from the impact of the accretion stream on the eccentric disc, which would release more energy when the stream falls furthest. This would occur

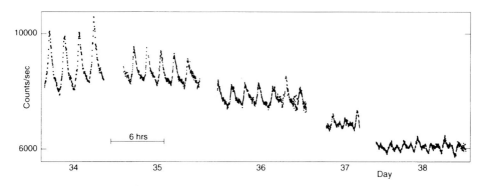

Fig. 6.8: The evolution of superhumps over five nights of the early decline of a superoutburst of V1159 Ori. Much space has been deleted between each data segment (add 244 9300 to the day to obtain JD). (Adapted from work by Joe Patterson and colleagues.[10])

84   Elliptical discs and superoutbursts   Ch 6

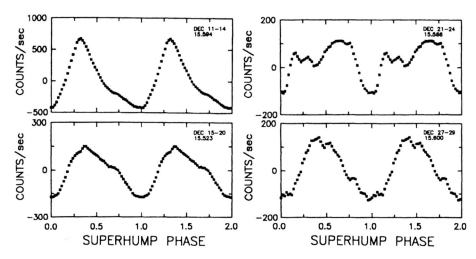

Fig. 6.9: Changing superhump waveforms during a superoutburst of V1159 Ori. (Figure by Joe Patterson.[10])

when the smaller side of the eccentric disc is nearest the secondary, which is the opposite to the situation for normal superhumps, and so would explain the half-cycle phase change.

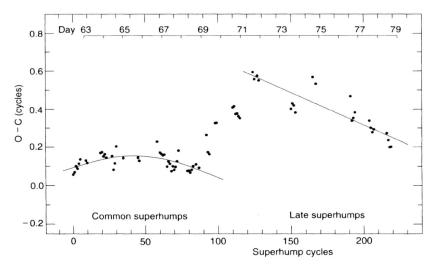

Fig. 6.10: The $O-C$ diagram for superhumps of IY UMa. Early on the superhumps show a trend to decreasing period (illustrated by the curved line); there is then a rapid change during which $O-C$ increases by half a cycle, so that the late superhumps are anti-phased with their predecessors. The period of the late superhumps is shorter than that of the common superhumps, as shown by the sloped, straight line. (Analysis by Joe Patterson using CBA data.[11])

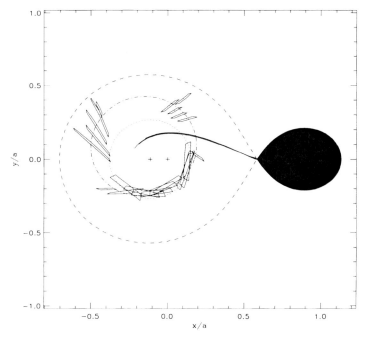

Fig. 6.11: By using the eclipses of the bright-spot (caused by the impact of the accretion stream with the disc edge) one can locate the edge of the disc. Eclipses at different superhump phases, with the disc at different orientations, then trace out the shape of the disc. In this figure, each rectangular region marks the constraint on the disc edge from an eclipse, derived from the phases of bright-spot ingress and egress and knowledge of the trajectory of the stream. Put together, the data indicate that the disc is elliptical. (Analysis by Dan Rolfe, using lightcurves of IY UMa during a superoutburst.[12])

Box 6.3: The $O - C$ diagram

The 'observed' minus 'calculated' or $O - C$ diagram for a periodic phenomenon — for instance the times of eclipse or times of hump maximum — is constructed using a test ephemeris. This ephemeris (the recurrence period together with the time of eclipse or maximum) is usually derived from earlier data. By assuming that the period does not change, one can extrapolate into the future to obtain a set of calculated times. One then plots the difference between the actual observed time and the calculated time (the $O - C$) to detect changes in the period. A horizontal line in an $O - C$ plot implies no change from the test ephemeris. A straight but sloped line implies a constant period, but one that is different from the test period. A curve in the $O - C$ plot implies a changing period: a deviation to larger $O - C$ means that the period is lengthening, delaying the observed times, whereas a decreasing $O - C$ implies a decreasing period.

6.4 THE SUPERCYCLE

As is obvious from Fig. 6.1, an SU UMa star undergoes a superoutburst, followed by a succession of regular outbursts, and then another superoutburst. The interval between superoutbursts — refered to as the *supercycle* — and the number of regular outbursts per supercycle, are characteristic of each star.

The reasons for this behaviour are still under debate, but a combination of the thermal instability and the tidal instability, as proposed by Yoji Osaki, is almost certainly involved.[13] Osaki's idea is summed up in the simulation shown in Fig. 6.12. The mass-transfer rate from the secondary is considered to be constant, but normal outbursts (which are thermal instabilities, as discussed in the last chapter) remove less material from the disc than has been fed into it since the last outburst. The disc thus gains mass and increases in size through a succession of such outbursts. Eventually, an outburst pushes the outer disc into the region where it is tidally unstable (remember that the enhanced viscosity of a thermal instability causes the disc to expand at the start of outburst). The disc becomes elliptical and starts to precess, initiating superhumps. The tidal stresses greatly increase the angular-momentum drain on the disc, enhancing the inward flow of matter, which sustains the disc in its hot state. Since the accretion rate now exceeds the rate of mass

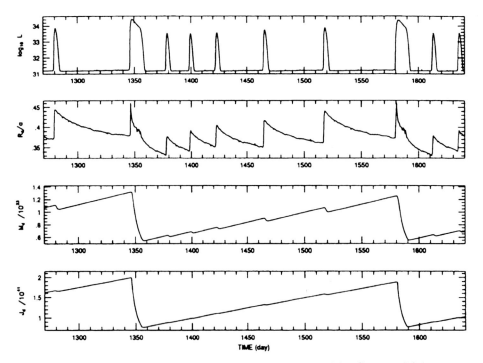

Fig. 6.12: Model simulations of a supercycle, computed by Susumu Ichikawa, Masahito Hirose and Yoji Osaki.[14] The panels show (from the top) the luminosity, the disc radius, the mass of the disc, and its angular momentum.

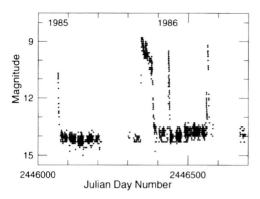

Fig. 6.13: In 1985 U Gem underwent an exceptionally long 45-day outburst, looking just like a superoutburst. Yet U Gem has too long a period to undergo the tidal instability. As pointed out by Jean-Marie Hameury and colleagues,[15] more material accretes during this outburst than was in the disc at its start, thus requiring that enhanced mass transfer plays a role in this outburst, and perhaps also in superoutbursts. (Data by the AAVSO.)

transfer, the disc begins to shrink.

Despite the fact that the disc becomes elliptical only when the disc expands beyond $r \sim 0.46a$, it can remain elliptical until it has shrunk to $r \sim 0.35a$. The tidal stresses thus drain the disc to this radius, by which time most of its mass has accreted [remember that surface density is higher at larger radii (Fig. 5.8) and also that there is more surface area at large radii]. The superoutbursts thus last much longer than normal outbursts, since they involve a much greater fraction of the disc's mass. After the superoutburst, the disc begins to accumulate mass again.

The tidal-thermal-instability model explains many features well, such as (1) the fact that superoutbursts are usually triggered by normal outbursts; (2) that the superoutbursts are regular (the supercycle length is set by the mass-transfer rate), and (3) that the spacing of normal outbursts increases through the supercycle (as is seen in Fig. 6.12 and also, though not always, in some observations). A further prediction of the tidal-thermal-instability model is that the disc expands through the supercycle, but we currently don't have the data to test whether this occurs.

6.4.1 Enhanced mass transfer?

Although the tidal-thermal-instability model provides a neat explanation of SU UMa stars, it may not be the whole story. There is evidence that the extra luminosity of outburst irradiates the secondary, and that this might lead to enhanced mass transfer. Such evidence includes the observed heating, in some outbursts, of the side of the secondary facing the white dwarf; increases in the 'orbital hump' during outburst, suggesting an enhanced mass-transfer stream, and anomalous outbursts such as that shown in Fig. 6.13, where the length of the outburst requires that enhanced mass transfer be involved. It is currently unclear how widespread or important such irradiation-induced mass transfer is.

6.5 ER UMA STARS AND WZ SGE STARS

The supercycle lengths of the known SU UMa stars are mostly around a few hundred days, but a few systems have much shorter or much longer supercycles. Since it

88 Elliptical discs and superoutbursts

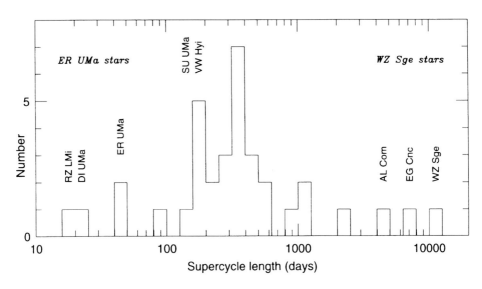

Fig. 6.14: The distribution of supercycle lengths amongst SU UMa stars. The shortest supercycle systems are refered to as ER UMa stars and the longest as WZ Sge stars.

is useful to have labels to refer to the extremes of SU UMa behaviour, the short-supercycle systems have become known as ER UMa stars,[†] while the long-supercycle systems are called WZ Sge stars.

ER UMa stars spend typically a third to half their time in superoutburst, with a supercycle of only 20–50 days. Outside superoutburst they pack in a rapid succession of normal outbursts, showing one every ~ 4 days (Fig. 6.15).

WZ Sge stars, by contrast, have supercycles lasting decades, while normal outburst are few and far between. WZ Sge itself has a superoutburst every 33 years and has never been seen to undergo a normal outburst.

The factor determining the different timescales appears to be mass-transfer rate. A typical SU UMa star has a mass-transfer rate of $\sim 5 \times 10^{12}$ kg s^{-1} (10^{-10} M$_\odot$ yr^{-1}); increase this to $\sim 4 \times 10^{13}$ kg s^{-1} and ER UMa behaviour results, the rapid succession of outbursts and superoutbursts being necessary to cope with the enhanced flow of material. Increase the mass transfer yet further, and the superoutbursts would lengthen, sustained by the extra material; with high enough mass transfer the star would become stuck in a permanent superoutburst and would be a novalike.

Our understanding of ER UMa behaviour, though, is not yet complete. One concern is why the superoutburst shuts off before it has drained the disc of so much material that the next normal outburst is delayed (whereas it actually occurs more or less as soon as the star reaches quiescence). In Osaki's standard tidal-thermal-instability model the enhanced mass flow through the disc while in its eccentric state ensures that an eccentric disc is always on the hot side of the thermal

---

[†]Or sometimes as RZ LMi stars.

instability. Thus the superoutburst ends when the disc reverts to being both cold and circular. However, this cannot apply to ER UMa stars, since superhumps are observed at nearly all phases of the supercycle, including during quiescence and in normal outbursts; thus the disc appears to be eccentric at all times.

Perhaps in ER UMa stars the dissipation in the eccentric state is too weak to sustain the disc in a hot state; in these stars the superoutburst ends with the disc dropping out of the hot state, while still eccentric. The eccentric dissipation is sufficient, though, to ensure that a new thermal outburst is triggered almost immediately, thus explaining the rapidity of the outburst cycle in ER UMa stars.

WZ Sge stars, by contrast, have a much lower mass-transfer rate, perhaps only 10^{12} kg s^{-1}. It then takes decades to accumulate sufficient material for a superoutburst. The puzzle of such stars, though, is why they show few or no normal outbursts during this interval. Even with a low mass-transfer rate, material should accumulate, drifting viscously into the inner disc, and trigger an outburst. One suggestion for why this does not occur is that the disc viscosity is very low, with $\alpha \lesssim 0.001$ (normally $\alpha \sim 0.01$–0.05 in a quiescent disc). The material would then remain in the outer disc, where much more can be stored before an outburst is triggered. The problem with this idea, however, is to explain the ultra-low viscosity.

An alternative explanation involves the removal of the inner disc, to prevent outbursts starting there. This could occur through 'siphons' or because of a magnetic field on the white dwarf. These possibilities are discussed in Chapters 7 and 9.

Another peculiarity of WZ Sge stars is the observation, early in superoutburst, of bright humps in the lightcurve, recurring with the orbital period, not the superhump period. In WZ Sge itself these persist for ~ 12 days before being replaced by superhumps. The origin of such humps is not understood. One possibility is that they come from the bright spot and indicate greatly enhanced mass transfer at the start of a superoutburst. However, this runs counter to the prevailing orthodoxy that dwarf-nova outbursts are not mass-transfer events. Another possibility is that

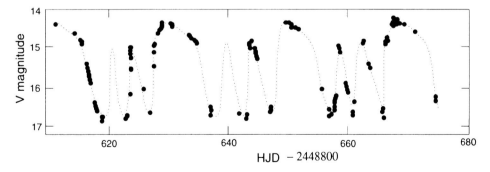

Fig. 6.15: The star with the shortest-known supercycle, RZ LMi, spends half its life in superoutburst, and has time for only two normal outbursts in between. (The 'Roboscope' data are by Robertson, Honeycutt and Turner[16]; the dotted line has been added to guide the eye.)

they arise from the early development of a disc eccentricity, before precession has set in.

6.5.1 EG Cnc and 'echo' outbursts

In 1996 the WZ Sge star EG Cnc entered superoutburst for the first time in 19 years.[17] In the month after the superoutburst it showed a succession of six 'echo' outbursts, after which it declined by a further magnitude into full quiescence, and no further outbursts were seen (Fig. 6.16).

Clearly, in the month after superoutburst, the mass flow through the disc was greater than in full quiescence. Why is this? One suggestion is that irradiation of the secondary during superoutburst heats the secondary, increasing the mass-transfer rate until it cools. However, it can also be explained by the idea discussed above for ER UMa stars. As in ER UMa stars, note that superhumps were seen in EG Cnc throughout the period of echo outbursts, indicating that the disc was eccentric throughout this time. If the end of the superoutburst is merely a reversion to a cold disc, the tidal dissipation then sustains a mass flow sufficient to drive the echo outbursts; eventually the drain of matter from the disc causes it revert to the circular state and the star drops to full quiescence.

If the above explanation is correct, why is the tidal dissipation insufficient to sustain the disc in a hot state in ER UMa and WZ Sge stars, when it appears to manage it in normal SU UMa stars? One clue is that both ER UMa and WZ Sge stars have the shortest orbital periods (all have $P_{\text{orb}} <$ 92 mins, whereas SU UMa stars range up to 170 mins) and so will have the lightest secondary stars and the smallest mass ratios. Indeed, some WZ Sge stars may be evolving through the period minimum, where the mass of the secondary star drops dramatically.

The smaller the mass ratio, the more room there is for the disc to grow at radii *outside* the 3:1 resonance with the secondary-star orbit, where dissipation is reduced. Thus, in the shortest-period systems, the weakened tidal dissipation might allow a cooling wave to develop outside the 3:1 resonance radius, ending the superoutburst

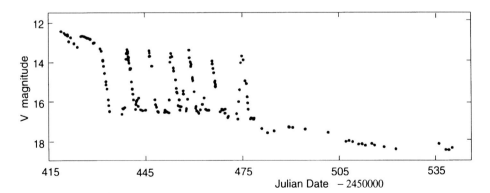

Fig. 6.16: After its first superoutburst in 19 years, EG Cnc showed a succession of six 'echo' outbursts, before dropping to full quiescence. (Data by the CBA.[17])

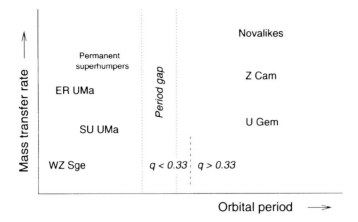

Fig. 6.17: Dwarf nova and novalike subtypes explained by varying two parameters, the mass-transfer rate and the orbital period (which is an easier-to-measure proxy for the mass ratio, q). (After a figure by Yoji Osaki.[13])

while the disc is still eccentric. When this happens, and if the mass flow through the disc (enhanced by the tidal dissipation) is less than the mass-transfer rate, the disc would fill up towards another superoutburst (ER UMa behaviour). If the mass flow in the disc were greater than the mass transfer, then echo outbursts would recur until the disc contracted to its circular state (EG Cnc behaviour).

6.6 PERMANENT SUPERHUMPS

Nothing in the tidal-instability model limits superhumps to appearing solely in outbursts. The only requirements are a sufficiently small mass ratio and a sufficiently large disc — conditions that can be satisfied by short-period novalikes, and then sustained by the high mass-transfer rate. Observers — notably the Center for Backyard Astrophysics — have looked for such *permanent superhumps* and have found them in nearly all novalikes with periods $P_{\rm orb} < 3$ hrs and in many with periods of 3–4 hrs.[18] Furthermore, the superhump period-excess, ϵ, obeys the same ϵ–$P_{\rm orb}$ relation as superhumps in SU UMa stars (they are both plotted in Fig. 6.6).

The finding of permanent superhumps has allowed a 'unification' of a wide sweep of cataclysmic variable behaviour based on varying only two variables, the mass-transfer rate, $\dot{M}$, and the mass ratio, or its proxy, orbital period. Consider first the systems with $P_{\rm orb} > 3$–4 hrs, which are tidally stable (see Fig. 6.17). The variation in $\dot{M}$ then causes the sequence from dwarf novae through Z Cam stars to novalikes. Below the divide at 3–4 hrs, a parallel sequence of systems with tidally unstable discs runs from WZ Sge stars (lowest $\dot{M}$) through SU UMa stars to ER UMa stars and then to the novalikes showing permanent superhumps (highest $\dot{M}$).

Fig. 6.18: Illustrations of a tilted disc. Each drawing shows the binary at the same orbital phase while the tilt precesses through one cycle. The tilt precession is retrograde (clockwise, in this illustration, compared to the anti-clockwise orbital motion). Note that the visible area of the disc varies with the tilt precession.

6.7 NEGATIVE SUPERHUMPS, OR 'INFRAHUMPS'

Superhumps have periods a few percent longer than the orbital period. Some stars, though, show humps recurring with a period that is shorter than the orbital period. These are commonly refered to as *negative superhumps*,[18] although I will use the shorter and neater term 'infrahumps'.

The most promising explanation for such humps involves a tilted accretion disc. Recall that normal superhumps involve an elliptical disc, whose axis of elongation precesses. This is refered to as 'apsidal' precession, where the 'apsides' are the points of an orbit furthest from and nearest to the white dwarf. A disc which is tilted out of the plane will also precess. This is refered to as 'nodal' precession, where the 'nodes' are the points where the edges of the tilted disc pass through the orbital plane. Nodal precession is *retrograde* (in the opposite direction to the orbital motion). This results in interactions between the secondary and the tilted disc occuring more frequently than the orbital period, generating a faster 'infrahump' period.‡

As yet this explanation is speculative, since there is no direct evidence for tilted discs in cataclysmic variables, and neither is there a theoretical demonstration of their occurrence. One clue is that infrahumps occur only in novalikes, and are not seen in SU UMa stars. Perhaps this means that they originate from a much weaker instability that requires far longer to grow than is available during outburst.

Supporting evidence for the tilted-disc model comes from the observation, in systems showing infrahumps, of brightness variations over the (implied) nodal-precession period. A tilted disc will change its aspect to Earth over this period, being seen closer to face-on at one precessional phase and closer to edge-on half a precessional cycle later, thus changing its apparent brightness (see Fig. 6.18).

Note also that in systems where both superhumps and infrahumps occur, the nodal precession period (typically ~ 4 days) is about twice the apsidal precession period (typically ~ 2 days). This leads to infrahumps having a similar ϵ–$P_{\rm orb}$ relation to superhumps (Fig. 6.19).

---

‡If $P_{\rm ih}$ is the infrahump period and $P_{\rm nodal}$ is the nodal precession period, then (cf. Box 6.1) $1/P_{\rm ih} = 1/P_{\rm orb} + 1/P_{\rm nodal}$.

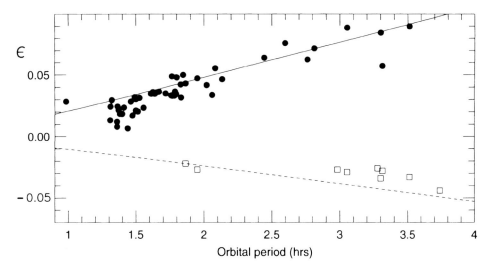

Fig. 6.19: The superhump period-excess, ϵ, for both superhumps (circles) and infrahumps (open squares).[7] The dashed line for infrahumps follows the relation $\epsilon_{ih} = -\frac{1}{2}\epsilon_{sh}$, which matches the data roughly (see also Fig. 6.6).

6.7.1 The problem of TV Col

There is one fly in the ointment of our understanding of eccentric discs. TV Col is a novalike with an orbital period of 5.5 hrs, where we would expect it to have a mass ratio of $q \sim 0.7$–0.8, far in excess of the theoretical limit for disc resonances. Yet TV Col's lightcurve shows infrahumps, the nodal precession period, and possibly also superhumps. Either TV Col is an exceptional system, having an anomalous mass ratio for its period, or the theory is lacking.[19] Note, though, that TV Col is also magnetic (see Chapter 9) and this might explain the discrepancy, since the effect of a magnetic field on the disc behaviour is uncertain. To investigate this issue observers need to search for superhumps in other stars with long orbital periods. To date, most searches have concentrated on shorter-period systems, where superhumps are expected, so we don't yet know whether TV Col is unique.

6.8 SPIRAL SHOCKS

This chapter has addressed the effect of the secondary on the disc. However, driving the disc elliptical, if the orbits resonate, is not the only result: the secondary can also trigger spiral shocks in the disc.

If there were no secondary, the gravitational field around the white dwarf would be symmetric, and thus the orbits in the disc would be circular. But the gravity of the secondary breaks the symmetry, producing non-circular orbits. Gas in the disc, orbiting much faster than the speed of sound, tries to follow the non-circular orbits, but in doing so forms shocks. Such shocks slow down the gas, converting its

94 Elliptical discs and superoutbursts   Ch 6

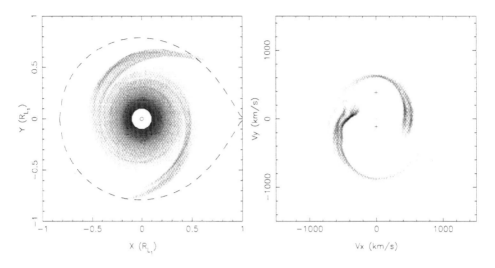

Fig. 6.20: *Left:* The spiral-shock structure induced by the secondary star, based on a hydrodynamical simulation of a disc in outburst. Darker colouring corresponds to greater line emission; the dashed line is the primary's Roche lobe, with the secondary to the right; axes are in units of the distance to L_1. *Right:* The Doppler tomogram resulting from the simulated line emission at left. The crosses mark the velocities of the white dwarf and secondary star. (Figures courtesy of Danny Steeghs.[20])

energy into turbulence and heat. For an analogy consider fast-moving traffic on a busy road. A simple lane change can force the car behind to brake, hence forcing the car behind that to brake harder, and so on. In heavy traffic such 'shocks' can bring traffic to a halt, even when the road is otherwise clear. The cars move slowly through the shock and then speed up again. If lower speeds are enforced by speed restrictions, and if lane swapping is discouraged, no shocks form and the overall flow of traffic can be increased.

In accretion discs, shocks have long been expected for this reason, and they have appeared in computer simulations which model discs using the equations of gas hydrodynamics. The usual result is a two-armed, spiral-shaped shock (Fig. 6.20). The spiral shape arises since the faster orbits in the inner disc 'wind up' the shock. This is reminiscent of spiral structure seen in many galaxies. Indeed, the spiral arms in galaxies may be shocks induced by interactions with neighbouring galaxies, which perturb the galactic orbits in the same way that the secondary perturbs the disc.

It was only in 1997 that observations of spiral shocks in cataclysmic variables were first reported. Danny Steeghs and colleagues analysed the emission lines of IP Peg near the peak of an outburst,[21] and found variations in the line profiles which matched those expected from a spiral structure. Indeed, the spiral structure became apparent when analysed by Doppler tomography (Fig. 6.21).

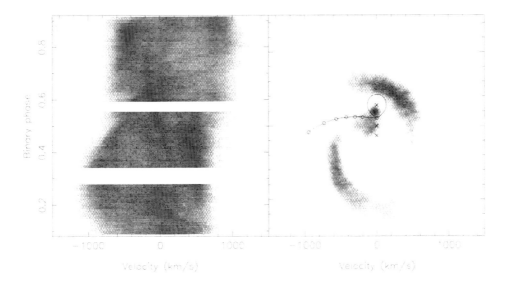

Fig. 6.21: *Left:* Phase-resolved line profiles of the He II λ4686 line of IP Peg during outburst (darker colouring corresponds to greater line emission; the white stripes are data gaps). *Right:* The Doppler tomogram of the data. The two-armed spiral structure can be compared to the tomogram of simulated data in Fig. 6.20. There is also emission from the secondary star (the secondary's Roche lobe is shown along with the accretion stream from L_1; crosses mark the system's centre of mass and the centroids of the two stars). (Figures courtesy of Danny Steeghs.[22])

At the time of writing, spirals shocks have been convincingly demonstrated in only one star, IP Peg, and then only during outburst. Thus it is currently unclear how widespread or important such shocks are. It may be that the shocks in many cataclysmics are too weak to be significant, or it may be that spiral shocks play a vital role in allowing material to flow through a disc. The shocks, through slowing material down, transfer angular momentum outwards through the disc, and so might drive accretion in the absence of other forms of viscosity. However, theory suggests that spiral shocks create an effective viscosity of only $\alpha \sim 10^{-4}$–10^{-2}, which is insufficient for them to be the sole contributor to viscosity.

Chapter 7

Siphons, winds and streams

The last two chapters discussed why discs undergo outbursts and why they can become elliptical. To complete the picture, we need to pay more attention to how material enters the disc (the stream–disc interaction) and how it leaves the disc (the boundary layer).

7.1 THE BOUNDARY LAYER

The Keplerian velocity just above the white dwarf surface is ~ 3000 km s^{-1}, yet the white dwarf is typically spinning much more slowly, with a surface speed of only ~ 300 km s^{-1}. The accreting material must therefore be slowed down to the speed of the white dwarf in a transition region known as the *boundary layer* (Figs 7.1 and 7.2). The kinetic energy of the slowing material is converted to heat and radiated away; the hot boundary layer can thus emit up to half the total luminosity.[1]

At high accretion rates the boundary layer contains enough material to block the emerging radiation, making it optically thick. The whole layer heats up until it emits blackbody radiation at a characteristic temperature of $\sim 200\,000$ K, roughly six times hotter than the hottest part of the disc.

7.1.1 Siphons

When the accretion rate is very low, less than $\sim 3 \times 10^{13}$ kg s^{-1} (5×10^{10} M$_\odot$ yr^{-1}), the above picture is altered. Very tenuous material at a high temperature finds it difficult to radiate energy and cool down. The dominant form of cooling is collisions between charged particles, leading to 'bremsstrahlung' radiation. But in low-density material there are few collisions. Being unable to cool, the hot gas expands, lowering its density further, making it even less able to cool, and so expanding further. The inner disc thus evaporates into a hot, diffuse 'corona' (Fig. 7.3). Effectively, the boundary layer is now blown up into a much larger but tenuous and optically thin region with a temperature of $\sim 10^8$ K (see Box 7.1).[2]

Such a corona supports itself by a 'siphon' effect. The flow of material towards the white dwarf releases the gravitational energy to keep the corona hot, and the

98 Siphons, winds and streams

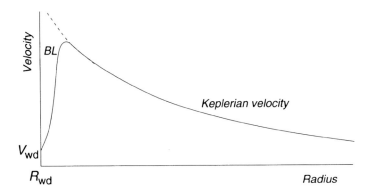

Fig. 7.1: The orbital velocity of material in the disc increases at smaller radii ($v \propto 1/\sqrt{r}$) until in the boundary layer (BL) the velocity plummets in order to match the much smaller surface velocity of the white dwarf.

hot corona conducts energy into the disc, evaporating material to replenish the corona. To conserve angular momentum, a portion of the corona flows outward, and eventually condenses in the outer disc.

Several observed phenomena can be explained by invoking a siphon. For instance, dwarf nova are observed to emit X-rays more strongly in quiescence than in outburst. The quiescent emission arises from the corona, which produces a small number of very energetic X-rays. In outburst, the accretion rate climbs by an or-

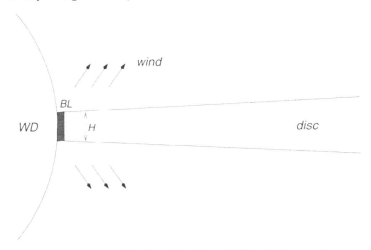

Fig. 7.2: An edge-on view of the boundary layer (BL) at high accretion rates. The height of the disc (H) is only $\sim 0.01\, R_{\rm wd}$, and the radial extent of the boundary layer is smaller still. This small strip around the white dwarf can emit up to half the total luminosity of the system. A wind outflow is driven by radiation from the boundary layer and/or the inner disc.

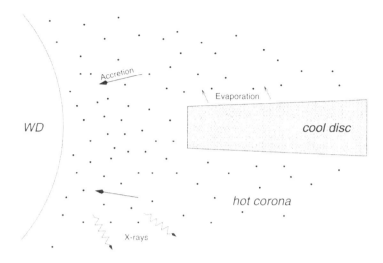

Fig. 7.3: At low accretion rates the boundary layer evaporates into a hot, diffuse corona, giving rise to energetic X-rays. Note that accretion occurs over a much larger region than at higher accretion rates. (After a diagram by Friedrich Meyer.[2])

der of magnitude, so that the boundary layer can cool efficiently and revert to the optically thick case. This leads to much more copious but much cooler blackbody emission, peaking in the extreme ultraviolet (see Fig. 7.4).

Secondly, the coronal material flowing outwards over the main body of the disc will produce emission lines, thus explaining why the spectra of quiescent dwarf novae show strong emission lines even if the main body of the disc is optically thick.

The evaporation of the central disc into a corona can also explain the otherwise-puzzling 'UV delay' in dwarf nova outbursts. At the start of such outbursts, the steep rise in the UV flux takes place ~ 1 day after the optical flux has risen. This is explained if the heating wave quickly sends the outer (optically emitting) disc into outburst, but must wait to re-fill the central hole with material before the innermost (UV emitting) regions participate in the outburst.

The removal of the inner disc might also explain why WZ Sge shows no normal outbursts (Section 6.5), since they cannot originate in an inner disc that isn't there!

7.2 WINDS

An atom near the boundary layer will be pulled towards the white dwarf by gravity. But it will also be bombarded by energetic photons flooding outwards from the boundary layer, pushing in the opposite direction. If the atom absorbs enough photons it will be driven off, to become part of a 'wind' flowing out of the binary.[3]

Atoms are particularly efficient at absorbing photons that have exactly the right energy to raise an electron from its lowest-energy orbit (the ground state) to the next-lowest orbit (the first excited state). An atom in this excited state will quickly de-excite, reverting back to the ground state by emitting a copy of the absorbed photon. The atom can then absorb another photon, repeating the cycle. The rapid succession of excitation and de-excitation between these states is called a 'resonance transition', and the resulting spectral line is called a 'resonance line'.

Since the absorbed photons are travelling outwards from the boundary layer, their momentum will push the atom outwards; however, the de-excitation photons are emitted in random directions, so that their momentum has no net effect. Effectively, the atom is scattering photons out of the radiation outflow, and is being driven outwards as a consequence. This is the driving mechanism for the wind.

Box 7.1: The temperature of the boundary layer

The released gravitational potential energy at a distance r from a white dwarf of mass M is given by $U \sim GMm/r$, while with a Keplerian speed of $v = \sqrt{GM/r}$ the kinetic energy, $\frac{1}{2}mv^2$, is $\sim GMm/2r$ or $\sim \frac{1}{2}U$.

In the optically thin case the material collides with the white dwarf surface (or a shock) and converts the energy to heat. Equating the energy of a gas particle, $\frac{3}{2}kT$, with the kinetic energy gives a boundary-layer temperature of

$$kT \sim \frac{GMm_{\rm p}}{6R_{\rm wd}}$$

(where I've taken the average particle mass as half the mass of a proton, $m_{\rm p}$, since the hot plasma contains roughly equal numbers of electrons and protons). For a typical white dwarf mass and radius (0.7 M$_\odot$ and 8×10^6 m) the temperature is $\sim 2\times10^8$ K, or equivalently ~ 20 keV.

In the optically thick case, the same energy emerges as blackbody radiation after thermalising throughout the boundary layer. The boundary layer (Fig. 7.2) can be considered as a strip around the white dwarf with a height equal to the disc thickness near the white dwarf, H. Thus the emitting area is $\sim 2\pi RH$ and hence, equating the blackbody radiation to the rate of potential energy release,

$$2\pi R H \sigma T^4 = \frac{GM\dot{m}}{2R_{\rm wd}}$$

for an accretion rate $\dot{m}$. For typical values of $H \sim 0.01\,R$ and $\dot{m} \sim 10^{14}$ kg s^{-1} the temperature is $\sim 200\,000$ K, or equivalently ~ 20 eV.

Thus the boundary layer at high accretion rates (optically thick conditions) is a copious emitter of UV and soft-X-ray radiation, while at lower accretion rates (optically thin conditions) there is less emission, but that emission is harder X-rays.

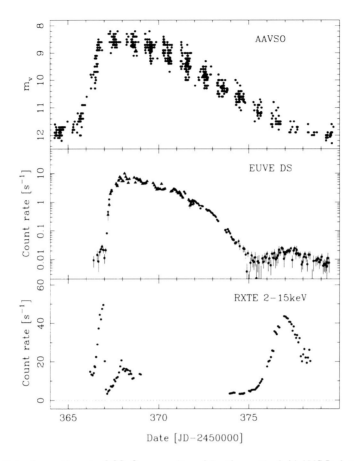

Fig. 7.4: An outburst of SS Cyg monitored in the optical (AAVSO data in top panel), in the extreme ultraviolet (*EUVE* Deep Spectrometer data in middle panel), and in the X-ray band (*RXTE* data in bottom panel). The outburst starts in the outer disc, so the optical brightness increases first. As increased accretion flows through the optically thin boundary layer the X-ray flux increases. Then, when the accretion rate is high enough that the boundary layer becomes optically thick, the X-rays are suddenly quenched, and the energy emerges instead in the extreme ultraviolet. The delay before the boundary layer becomes optically thick means that the ultraviolet brightness increases ~ 1 day after the optical rise. At the end of the outburst, when the accretion rate has dropped, the boundary layer again becomes optically thin and the X-ray flux rises. (Figure by Peter Wheatley, in collaboration with Christopher Mauche and Janet Mattei.[4])

7.2.1 P Cygni profiles

We know that cataclysmic variables have wind outflows — at least some of the time — since their resonance lines show the 'P Cygni profiles' characteristic of a radiation-driven wind. Consider the schematic outflow from a white dwarf drawn

102 Siphons, winds and streams   Ch 7

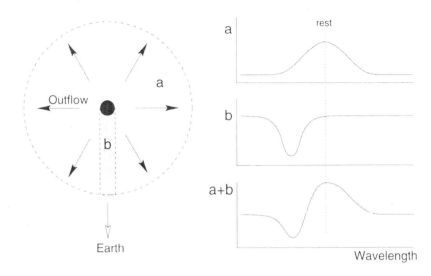

Fig. 7.5: A wind outflow from a white dwarf giving rise to P Cygni line profiles. The bulk of the outflow (a) emits scattered resonance-line photons, with a Doppler-broadened profile owing to the wind's motion. From the white dwarf (b) we see a hot continuum minus the photons scattered by the wind (which are blueward of the line's rest wavelength due to the wind's motion towards us). The overall line profile thus contains the mixture of line emission and blue-shifted absorption characteristic of outflows.

in Fig. 7.5. Some of the resonance-line photons are scattered towards us by the wind. The resulting line is broadened by Doppler shifts, since the flow ranges from directly away from us (red-shifted) to nearly towards us (blue-shifted). The wind is tenuous, and so there is little continuum emission from it.

Looking directly at the white dwarf, however, we see the strong continuum from the hot boundary layer. The continuum photons with the right energy to excite the wind atoms will be scattered, resulting in an absorption feature where they are missing. Since the wind in front of the white dwarf is flowing towards us, this absorption feature will be blue-shifted, by an amount comparable to the escape velocity.* Of course, we cannot resolve the different components spatially, so what we observe is the sum of the two components — a blue-shifted absorption trough cut into an emission line, referred to as a P Cygni profile.

7.2.2 Winds in cataclysmic variables

Judging by the presence of P Cygni profiles, cataclysmics possess winds when the accretion rate is highest, either in a novalike state or during the peak of outburst.

---

*The 'escape velocity' is the velocity that the wind must have to overcome the gravitational potential and escape from the binary. This is found by equating the kinetic and potential energies at the white-dwarf surface, $\frac{1}{2}mv^2 = GM_{\rm wd}m/R_{\rm wd}$, to obtain $v_{\rm escape} = \sqrt{2GM_{\rm wd}/R_{\rm wd}}$.

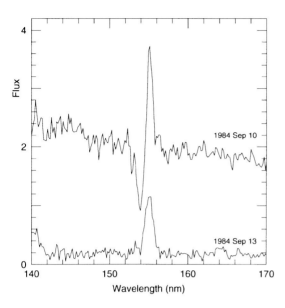

Fig. 7.6: Spectra of WX Hyi taken with the *International Ultraviolet Explorer*, covering part of the ultraviolet including the resonance line of C IV at 155 nm (1550Å). The first spectrum, taken at the peak of outburst, shows a classic P Cygni profile. The second, taken three days later during the outburst decline, shows no blue-shifted absorption.[5]

For instance, Fig. 7.6 shows spectra of WX Hyi at the peak of outburst, when P Cygni profiles were clearly present, and three days later on the outburst decline, by which time the P Cygni profiles had disappeared.

Our understanding of winds, however, is still at a rudimentary level. Amongst the issues still to be settled are: (1) Does the wind originate in the boundary layer or in the inner disc? The boundary layer is brighter, but the higher orbital velocity in the inner disc might help to accelerate the wind. Perhaps magnetic fields in the disc help to sling-shot particles outward. (2) How bipolar is the wind? — meaning how does the wind density depend on the angle above the orbital plane? (3) How does the wind depend on orbital phase? There are some hints that the P Cygni profiles vary with orbital phase in some systems, but there is as yet no good model for this. (4) How much of the accretion flow is driven off as a wind? Estimates have varied from a few per cent up to a less credible 30 per cent.

7.3 THE DISC–STREAM IMPACT

There is currently much research into the nature of the disc–stream collision, both observationally and by modelling the interaction on computers. The orbital hump and emission-line bright spot indicate that much of the kinetic energy of the stream is dissipated in the impact, but the details depend on uncertain parameters such as the height of the disc edge.

Observationally, we know that part of the material can be thrown upwards, out of the plane. This is deduced from X-ray lightcurves of high-inclination systems which show *dips* at orbital phase ~ 0.8. At this phase the stream–disc splash region is in front of the white dwarf, and material splashed out of the plane can absorb the energetic X-ray emission from the white dwarf and the boundary layer. We see such dips in U Gem,[6] and since we can estimate the system inclination from the grazing eclipse in this star (giving $i \approx 70°$), we deduce that in order to obscure the white dwarf the splash must extend to at least $\approx 20°$ above the plane (measuring the angle at the white dwarf). In FO Aqr a blue-shifted absorption feature in the emission lines appears, similarly, to come from material being splashed out of the plane.[7]

Although such dips occur preferentially, as expected, at phase ~ 0.8, they can also occur elsewhere, such as after eclipse (phases 0.0–0.2), where they are harder to explain (see Fig. 7.7). Perhaps the stream collision is setting up 'standing waves' along the disc rim, in which material oscillates above and below the plane as it orbits. This is currently speculative, but some such mechanism is needed to explain the observations.

There are also indications that not all of the stream is stopped by the interaction at the edge of the disc. Simulations (see Fig. 7.8) suggest that $\sim 10\%$ of the stream flows over the disc, being slowed by $\sim 30\%$ in the process, but managing to continue on a roughly free-fall trajectory until it re-impacts the disc much nearer the white dwarf (although this description depends on assumptions in the modelling, and so is uncertain). Material thrown up by the re-impact might also be the cause of dipping activity at anomalous phases. Observational evidence for the stream overflowing the disc is seen mainly in the magnetic systems, which we discuss in Chapters 8 and 9, and in a group of novalikes which have become known as SW Sex stars, to which we now turn.

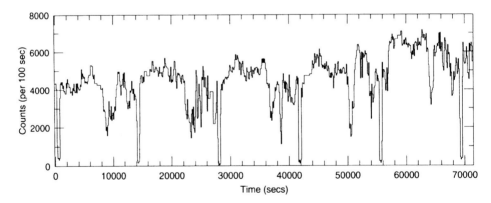

Fig. 7.7: This X-ray lightcurve of EXO 0748–676 shows deep eclipses recurring with the 3.82-hr orbital period, and obvious 'dipping' activity, presumably caused by material thrown out of the disc plane by the impact of the accretion stream with the disc. The dips are most prominent at phase 0.65, but occur throughout the range 0.60–0.15. Note also the marked variability from cycle to cycle, indicating that the stream–disc 'splash' is very chaotic. EXO 0748–676 is a low-mass X-ray binary, rather than a cataclysmic variable (which means it contains a neutron star rather than a white dwarf; see Chapter 13), but similar dipping activity occurs in high-inclination cataclysmics. (Data from the *EXOSAT* satellite; note that some 'bursts' have been edited out, for which see Section 11.5.)[8]

Fig. 7.8: A computer simulation in which particles representing an accretion stream (entering from the left) collide with a disc (seen edge on). Part of the stream is diverted out of the plane to flow over the disc. Note also the shock waves at the disc–stream impact. (By Philip Armitage and Mario Livio.[9])

7.4 THE SW SEX PHENOMENON

In coining the term 'SW Sex star' John Thorstensen drew attention to a group of novalike variables showing a set of peculiar characteristics that are not explained by the traditional picture of a novalike.[10] The first SW Sex stars identified were all eclipsing systems, and this has led to a debate about whether they are truly a distinct class, or whether the SW Sex characteristics are simply the appearance of novalikes when seen edge on. At the time of writing, this debate has not been settled, and no consensus about the cause of SW Sex characteristics has been reached. Thus the following account is open to revision.[11]

7.4.1 A flared disc?

An important clue to SW Sex stars was deduced by Christian Knigge and colleagues[12] when comparing UV spectra of DW UMa, obtained first when it was in its normal state, and then when it had faded by three magnitudes in optical light (the cause of such 'low states' is discussed in Chapter 12). During the low state they clearly saw the UV spectrum of a hot white dwarf, but in the high state there was no sign of the white dwarf, and the UV flux was actually lower. The implication is that the white dwarf in DW UMa is usually hidden, obscured by a thick, flared disc.†

If this is typical of all SW Sex stars, it implies that we are looking primarily at the cool outer wall of an edge-on, flared disc, and can see the hot, inner surface only on the far side, beyond the white dwarf (see Fig. 7.9). This can explain one of the SW Sex peculiarities, that eclipse mapping (see Box 3.2) produces results inconsistent with a flat, Keplerian disc. Instead of showing hotter regions nearer the white dwarf, eclipse mapping of SW Sex stars reveals a more uniform temperature, as expected from a disc rim.

---

†Since the inclination of DW UMa is $\approx 82°$ this requires that the disc is flared by at least $\approx 8°$.

Fig. 7.9: An illustration of a flared disc. For a disc flared by 10° above the orbital plane, the direct view of the white dwarf is obscured when the inclination is higher than 80°. In this case we see the cool outer rim of the disc predominantly, plus the hotter inner regions of the rear of the disc.

7.4.2 Stream–disc overflow?

A second SW Sex characteristic is seen in the wings of their emission lines, where a high-velocity feature zig-zags across the line profiles, moving with the orbital cycle, and extending right into the line wings (Fig. 7.10). If the effect of the stream's impact is confined to the low-velocity outer disc, we expect the bright-spot S-wave to be confined to the line core; the undisturbed, high-velocity inner disc would be symmetric about the white dwarf, and thus the line wings would move only with the orbital motion of the white dwarf. Clearly, either our view of the inner disc is being partially obscured by the rim of the flared disc, or the inner disc itself is being disturbed.

One proposed explanation is that part of the stream flows over the disc, perhaps punching a hole through the disc rim, and produces high-velocity line emission where it re-impacts the inner disc. Both the phasing and the velocity of the high-velocity feature are compatible with this idea (depending on how much the initial stream–disc impact deflects and slows the overflowing stream).

A third characteristic of SW Sex stars is deep absorption features seen in the cores of hydrogen and He I lines, typically confined to orbital phases ≈ 0.2–0.6 and deepest at phase ≈ 0.5 (thus leading to the shorthand 'phase 0.5 absorption'). It is possible that the absorption also results from the overflowing stream, which would be seen projected against the bright disc during the free-fall part of its trajectory, before the re-impact with the disc.[13]

Furthermore, since the disc is flared, so that only its rear is visible, the resulting absorption would be prominent only when the stream is on the rear of the disc, explaining why the absorption is seen only during phases 0.2–0.6. Note that the usual bright-spot S-wave is not seen in SW Sex stars, which would also be explained if this region were instead producing absorption.

7.4.3 Winds?

A fourth peculiarity of SW Sex stars is that one never sees the double-peaked lines expected from a Keplerian disc, even though they are high-inclination systems where double peaks should be seen most readily. This could be partly due to

Sec 7.4   The SW Sex phenomenon   107

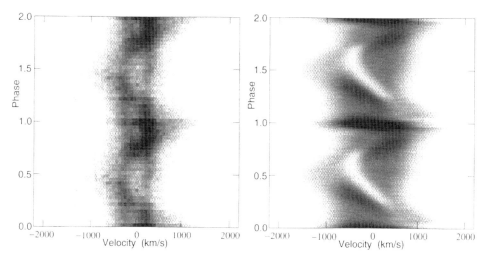

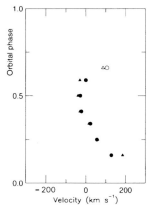

Fig. 7.10: *Top left:* The profile of the Hα line of the SW Sex star PX And varying over the orbital cycle (compare with WZ Sge's spectra in Fig. 3.8). An obvious high-velocity component zig-zags into the line wings. *Top right:* Simulated line profiles in which an overflowing stream gives rise to the high-velocity feature where it re-impacts the disc. Absorption in the line core from the ballistic part of the stream's path is restricted to phase 0.2–0.6 by a disc flare. The motion of the absorption from red to blue over this phase range matches the observed centroid of the absorption shown *right*.[13]

the disc flare, since a disc rim produces a flat-topped line profile, rather than a double peak. In addition, though, it appears that a narrower component from a disc wind also contributes, filling in the double peaks to produce a single-peaked profile overall. This is particularly so in the He II λ4686 line, which shows the orbital motion expected for an origin in the boundary layer and is dominated by the wind component. Furthermore, evidence for wind-formed P Cygni profiles has been seen in some SW Sex stars (Fig. 7.11).

7.4.4 Infrahumps?

Recently, it has become clear (mainly through the work of the Center for Backyard Astrophysics) that many SW Sex stars show infrahumps (negative superhumps).[14] This offers a possible explanation for stream overflow in these stars, since infrahumps most likely arise from a tilted disc, and the tilt would naturally allow the stream to flow above the disc at some precessional phases and below it at other phases. The obvious observational test — looking for changes in the emission lines over

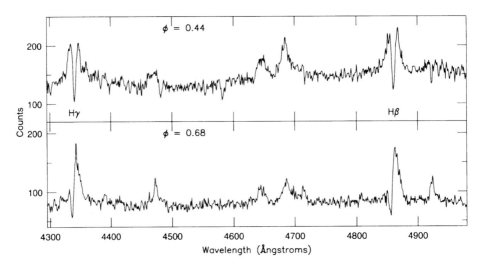

Fig. 7.11: Spectra of the SW Sex star V1315 Aql. At orbital phase 0.44 the Hβ and Hγ lines show deep 'phase 0.5 absorption' in their cores. The spectrum from phase 0.68 shows probable P Cygni profiles, indicative of a wind outflow.[15]

precessional phase — has not yet been pursued, although it is clear that SW Sex characteristics (phase 0.5 absorption and the high-velocity emission) are highly variable and thus that any overflow is variable and intermittent.

7.4.5 Relation to other novalikes

So are SW Sex stars a distinct class? It is clear that many of their properties, such as the obscuration by the rim of the flared disc, result from their high inclination. However, at least one system, LS Peg, appears to be a non-eclipsing SW Sex star.[16] Interestingly it shows He I absorption at nearly all orbital phases, as would be expected if the disc flare were less effective at confining it to the 0.2–0.6 phase range. However, more work is required to investigate SW Sex characteristics at a range of inclinations.

It is also unclear whether there is a clear divison between novalikes in which the stream overflows the disc, and those with no overflow. One possibility is that overflow occurs episodically in all novalikes, and that at such times they show SW Sex characteristics. Alternatively, perhaps the SW Sex stars are the novalikes with the highest mass-transfer rates, in which the high $\dot{M}$ results in one or more of tilted discs, disc flaring, overflow, and strong winds. It is noteworthy that SW Sex stars occur preferentially with orbital periods in the range 3–4 hrs, where there are other indications (notably a dearth of dwarf novae; see Fig. 5.15) that the mass-transfer rate of cataclysmic variables is increased. We return to this theme in Chapter 12.

Chapter 8

Magnetic cataclysmic variables I: AM Her stars

There is an apocryphal tale about an elderly astronomer who, at the end of any seminar by a young researcher, would always ask the same question: "Ah, but have you considered the effect of magnetic fields?" The point being that magnetic fields complicate a situation immensely, and make it much harder to address theoretically. So far this book has dealt with non-magnetic cataclysmic variables — those in which the magnetic field of the white dwarf is weak enough to have no effect on the accretion flow. But in systems with a strongly magnetic white dwarf the whole nature of the accretion process is changed radically.

8.1 MAGNETIC ACCRETION

A magnetic field affects the motion of charged particles, but moving charges generate magnetic fields. Thus any interaction of a field with a hot, ionised gas leads to a complex feed-back loop of interactions. Fortunately, the end result can be summarised by two general principles. The first principle is that the field and the matter become *frozen* together: the charged particles can move along field lines but cannot easily cross them, while any motion of material drags the field along with it.

The second principle is that, for the purposes of deducing the motion of the material, one can usually either ignore the field, or ignore everything but the field. Consider, first, accreting material far from a magnetic white dwarf. The kinetic energy associated with the bulk flow of the gas will far exceed the energy associated with the interaction with the field. The flow will thus continue as though there were no field (the field will be dragged along with the flow, but is too weak to affect it). Close to the magnetic white dwarf, however, the energy of the matter–field interaction greatly exceeds the energy of the bulk flow. In this regime the field lines remain immovable and the material can do nothing but flow along them.

Of course there will be a transition region between the two regimes, and that is the least understood part of a magnetic cataclysmic variable. However, since the strength of a magnetic field declines rapidly as one moves away from it, magnetic cataclysmic variables can often be regarded as having an outer zone, which acts much as it would in a non-magnetic system, and a magnetically dominated *magnetosphere* surrounding the white dwarf. This division is aided by an effect called *screening*. At the magnetospheric boundary the field induces electric currents in the ionised plasma; these currents counteract the effect of the field and so 'screen' the field from the material further out.

The extent of the magnetosphere is set mainly by the strength of the field and by the accretion rate (see Box 7.1), although it also depends on whether the material is confined to a stream or spreads out into a disc.

Inside the magnetosphere the material, locked to the field lines, is forced into corotation with the white dwarf — completing one orbit every white-dwarf spin period, regardless of its radius. This leads to a third general principle: the spin period of the white dwarf tends to adjust itself so that the circular motion just inside the magnetosphere equals the circular ($\sim$ Keplerian) motion of material just outside the magnetosphere. This is the equilibrium situation for which there is no large jump in velocity at the magnetospheric boundary. Of course there is no guarantee that a particular system will be in equilibrium; for instance a change in mass-transfer rate can alter the radius of the magnetosphere, but the inertia of the white dwarf will prevent its spin rate from adjusting to match, except on much longer timescales. Nevertheless, this principle means that the lowest-field white dwarfs will have the smallest magnetospheres, and will thus be spinning fastest; the higher-field white dwarfs will spin more slowly, finding an equilibrium with the slower-moving outer parts of the binary. Thus the interaction between field strength, spin rate and mass-transfer rate determines the subclass of magnetic cataclysmic variable.

8.2 THE HIGHEST-FIELD SYSTEMS

The white dwarf in AR UMa has the strongest field known among cataclysmic variables, measuring 230 MGauss (23 000 Tesla) at its surface[1] (which is thousands of times stronger than we can create in laboratories on Earth). This strong field interacts with the smaller magnetic field of the secondary and locks the two stars together so that they always present the same face to each other. Thus the white dwarf spins at the same rate as the two stars orbit — a *synchronous rotation* that is the defining characteristic of an AM Her star.[2]

The 230-MG field is so strong that it still dominates at the L_1 point, and the accretion stream is forced to follow field lines almost from the start (Fig. 8.1). The field lines form a dipolar pattern — the pattern obtained by scattering iron fillings around a bar magnet.* Thus to follow a field line the stream must divert out of the orbital plane. The stream splits into two, one part heading for the 'north' magnetic pole and the other for the 'south' pole. The field lines converge as they approach the white dwarf, squeezing the streams (flux freezing again) and funnelling them

---

*More complex fields are possible, such as a quadrupole field or a dipole field that is not centred at the white dwarf's centre, but for most purposes a dipole is an adequate description.

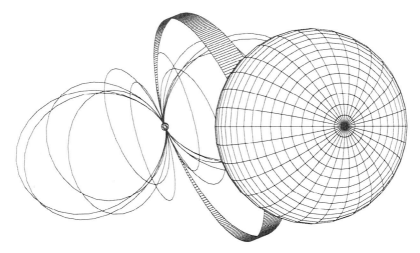

Fig. 8.1: In AR UMa the accretion stream emerges from the L_1 point, splits into two, and follows field lines onto the magnetic poles of the white dwarf. The dipole pattern of the field lines is shown. (Illustration by Gary Schmidt.[1])

onto tiny *accretion spots* near the poles, whose radii are only $\sim 1/100^{\text{th}}$ that of the white dwarf.

Being channelled by the field, the stream moves almost radially towards the white dwarf, in virtual free-fall. The potential energy is converted to kinetic energy (not dissipated by viscous interactions, as in a disc) and the stream slams into the white dwarf at roughly the ~ 3000 km s^{-1} escape velocity. In the resulting *accretion shock* the kinetic energy is converted into X-rays and radiated away. (In laboratories we generate X-rays by accelerating electrons in an electric field and slamming them into a metal target; this is the same principle, but the acceleration is due to gravity.) Magnetic cataclysmic variables are thus stronger X-ray sources than their non-magnetic counterparts, emitting most of their energy as X-rays and extreme-ultraviolet photons.

8.3 THE ACCRETION STREAM IN AM HER STARS

Whereas AR UMa's 230-MG field controls the stream from the L_1 point, in AM Her stars with more typical fields (10–80 MG) the stream is at first unaffected by the field, and flows on a 'ballistic' trajectory (as it would in a non-magnetic system) until nearer the white dwarf. Provided the magnetosphere extends out further than the circularisation radius (see Section 2.4), the stream cannot orbit freely and so does not form a disc. Where the magnetic field begins to dominate, the stream changes direction, and must flow out of the plane to follow field lines for the rest of its journey (Fig. 8.2).

The details of the transition from ballistic to magnetically controlled flow are complex. As the stream approaches the white dwarf, the increasing magnetic pres-

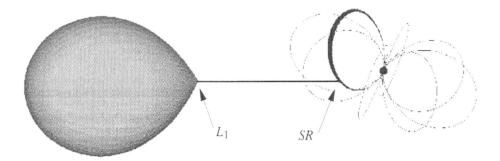

Fig. 8.2: A schematic AM Her star showing the ballistic portion of the stream between the L_1 point and the stagnation region, SR, followed by the flow onto the magnetic poles. Only the field lines corresponding to the magnetospheric radius are drawn. (Illustration by Jens Kube.[3])

sure of the converging field lines first squeezes the stream, causing it to break up into dense 'blobs' of material. The field cannot easily penetrate such blobs because of screening, so they continue ballistically for a while. As the magnetic pressure climbs the blobs are forced to change direction; collisions in the stream form shocks, energy is dissipated and radiated away, and a pool of material can collect in a 'stagnation' or 'threading' region. This region extends over a range of radii, owing to the range of blob densities. Material from the pool — a mixture of blobs and a fine 'mist' of material stripped from the surface of the blobs — then diverts along field lines and flows onto the white dwarf.

Physical systems tend to settle into their lowest-energy configuration, but diverting the stream out of the plane requires energy. To minimise this, it is found that in many systems the magnetic field of the white dwarf has tilted over, so that one magnetic pole 'points' towards the direction from which the stream comes (Fig. 8.2). As a result, material flows preferentially onto that pole; material can still flow to the other pole, but only by going the long way round, moving further out of the plane, and only a small fraction succeeds in doing this. If the dipole is offset from the white dwarf centre, one magnetic pole will be stronger than the other. In this case the stream will prefer to feed the weaker pole, since this again requires less of a diversion.

Eclipses in high-inclination AM Her systems provide a graphic illustration of the stream geometry. Fig. 8.3 shows an eclipse of HU Aqr, along with an illustration of the system approaching eclipse. The lightcurve reveals that the tiny, and thus rapidly eclipsed, accretion spot at the magnetic pole emits roughly half the total light; the other half comes from the extended stream, which enters and leaves eclipse more gradually.

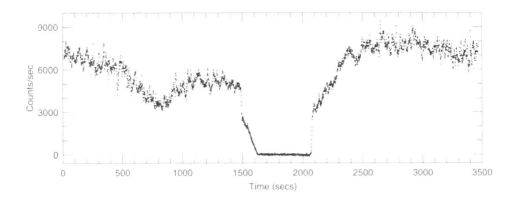

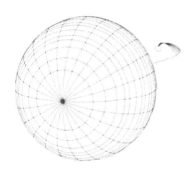

Fig. 8.3: An optical lightcurve of the AM Her star HU Aqr (*above*) and an illustration of the system approaching eclipse (*left*), corresponding to 1300 sec in the above plot.[4] At 1490 sec the tiny accretion spot is eclipsed, and the light drops dramatically. The bright stream enters eclipse over the next 130 sec. At the end of the eclipse the accretion spot suddenly emerges from behind the red dwarf (2075 sec), followed by the stream (2075–2300 sec). Earlier, at around 800 sec, the stream had been in front of the accretion spot, absorbing some of its light and causing a dip in the lightcurve.

8.4 AM HER X-RAY LIGHTCURVES

The accretion pattern in AM Her stars, funnelling mainly onto a small spot near one magnetic pole (with some material possibly flowing to a similar spot at the opposite pole), leads to characteristic X-ray lightcurves. First, consider a system in which accretion is entirely onto one pole (such as that illustrated in Fig. 8.3). When that pole is on the visible face of the white dwarf we see plenty of X-rays, but while it is on the far side of the white dwarf we see none. The lightcurve looks like that of ST LMi or VV Pup (see Fig. 8.5), with near-zero flux for roughly half the cycle.

If the system were seen at a low inclination, the accreting pole might always be on the visible face.[†] Thus the X-ray flux would be roughly constant with orbital phase. (Alternatively, if accretion went to the lower pole, it might never be visible, and no X-rays would be seen.) Some of these systems do, though, show 'dips' in the X-ray lightcurves, and these result from the accretion stream temporarily obscuring

---

[†]For an inclination i, and denoting the angle between the rotation axis and the magnetic axis by β, and the magnetic colatitude of accretion by ϵ, this occurs if $i + \beta + \epsilon < 90°$.

Box 8.1: Magnetic accretion

The strength of a magnetic dipole is expressed as a 'magnetic moment', μ. In AM Her stars the magnetic moments of the white dwarfs are typically $\mu = 10^{24}$ T m^3 (or 10^{34} G cm^3). The field strength, B, declines as μ/r^3 so that at the white dwarf surface ($r \approx 7 \times 10^6$ m) B is of order 3000 T (30 MG).

The dipole shape of the field lines is described by the equation

$$r = C \sin^2 \theta$$

where r is the distance to a field line at an angle θ to the magnetic axis and C is a constant. Choosing a different C specifies a different field line, emerging from the white dwarf at a different latitude (field lines do not change with the azimuthal angle ϕ).

The magnetic pressure is given by $B^2/2\mu_0$ (or $B^2/8\pi$ in cgs units) and thus falls off as r^{-6}, justifying the crude division into magnetospheric and ballistic regions. To deduce the radius at which the magnetic field takes over we compare this pressure with the pressure exerted by the infalling stream (the 'ram' pressure). Taking a cross-section of area A of a stream flowing at velocity v, the volume passing a point in time t is vAt; the mass is thus $\rho v A t$ and the momentum $\rho v^2 A t$. Dividing by time to get force and then by area to get pressure leads to the ram pressure ρv^2. Hence the magnetosphere begins where

$$\frac{B^2}{2\mu_0} = \frac{\mu^2}{2\mu_0 r^6} = \rho v^2.$$

Thus the transition to magnetic trajectories occurs closer to the white dwarf for denser material. The actual location will depend on how the material is distributed. For instance, if the accretion flow $\dot{M}$ occurs in a stream of cross-section A, one can write $\rho = \dot{M}/Av$. The stream velocity, assuming that it is in free-fall from a large distance, is found by equating kinetic energy and potential energy to get $v^2 \approx 2GM_{\rm wd}/r$. Combining the last three equations then gives

$$r_{\rm mag} = \left[\frac{A^2}{8GM_{\rm wd}} \frac{\mu^4}{\mu_0^2} \frac{1}{\dot{M}^2} \right]^{1/11}.$$

If the material forms a disc, which is then magnetically disrupted, this occurs at

$$r_{\rm mag} \approx 0.5 \left[\frac{1}{8GM_{\rm wd}} \frac{\mu^4}{\mu_0^2} \frac{1}{\dot{M}^2} \right]^{1/7}.$$

Knowing the radius of the magnetosphere, we can ask what its rotation would be if it were in equilibrium with its surroundings. If the white dwarf (and hence the whole magnetosphere) is spinning at $P_{\rm spin}$, then at radius $r_{\rm mag}$ the velocity of rotation is $2\pi r_{\rm mag}/P_{\rm spin}$. Setting this equal to the Keplerian velocity at the same radius (see Box 2.1) leads to

$$P_{\rm spin} = 2\pi r_{\rm mag}^{3/2} \, (GM_{\rm wd})^{1/2}.$$

Thus white dwarfs with larger fields (larger magnetospheres) tend to spin more slowly.

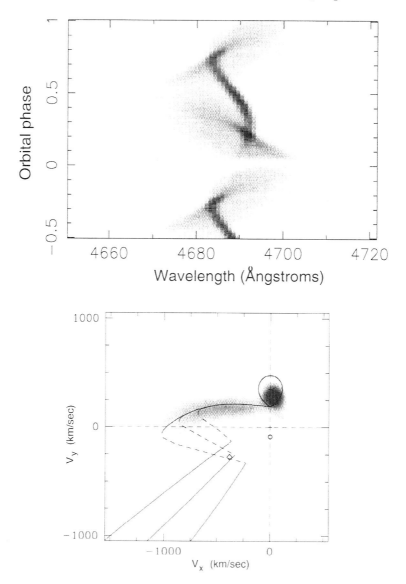

Fig. 8.4: In the absence of a disc, the line profiles of AM Her stars are dominated by contributions from the accretion stream and the irradiated face of the secondary. The *top* panel shows the line profiles of the He II $\lambda 4686$ line of HU Aqr, shown as a greyscale 'trailed spectrum' over the orbital cycle. Note the eclipse at phase 0, when little light is seen. The *bottom* panel is a tomogram of the data, overplotted with the secondary's Roche lobe and the trajectory of the stream, which has been calculated for three different locations of the threading region. The tomogram reveals that the brightest emission (the lower-velocity S-wave in the trailed spectra) arises from the illuminated secondary. The higher-velocity emission comes from the stream. (Figures by Axel Schwope and colleagues.[5])

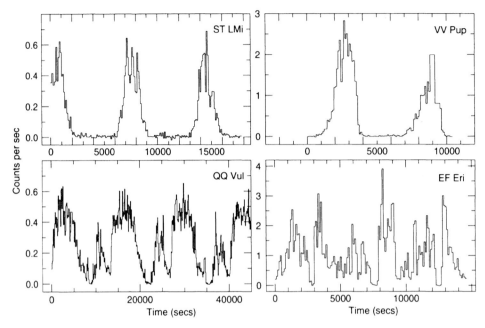

Fig. 8.5: X-ray lightcurves of AM Her stars. In ST LMi and VV Pup the main accreting pole disappears over the white dwarf limb for ∼ half the orbital cycle; in QQ Vul and EF Eri the accreting pole is always visible, though there are periodic dips when the accretion stream passes in front of the pole, absorbing the X-rays. The 0.04–2-keV lightcurves, covering 2–3 orbital cycles of each star, were observed with the *EXOSAT* satellite.[6]

the view to the accretion spot.[‡] The two types of behaviour are sometimes referred to as 'one-pole' behaviour (one pole is always visible) and 'two-pole' behaviour (we see alternately one pole and then the other).

The 'two-pole' behaviour is further complicated if some accretion passes to both poles. Occultation at the two poles will be ∼ 180° out of phase, and thus the gaps in the X-ray flux (when the dominant pole disappears) might be partially filled in by weaker X-ray emission from the second pole. We will return to this topic after having discussed the accretion polecap in more detail (Section 8.6).

8.5 CYCLOTRON EMISSION

The ionised material in an AM Her accretion stream does not simply follow a field line; instead it must spiral around the field line (Fig. 8.6). This arises for the following reasons. The motion of the charged particles in the stream is effectively an electric current. But any electric current moving perpendicularly to a field line will experience a force, and this force is perpendicular to both the field line and the direction of motion. Since the force is always perpendicular to the direction of

---

[‡]This occurs if $\beta \lesssim i$.

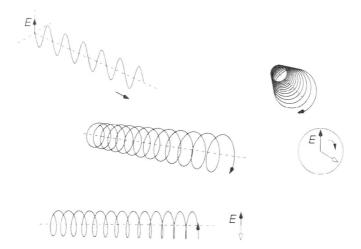

Fig. 8.6: The electric vector of a photon is perpendicular to its motion (*top left*). An electron following a field line spirals at the cyclotron frequency, and so produces cyclotron emission. When the line is seen side on (*bottom*) the apparent motion is predominantly up and down, producing photons that are linearly polarised in this direction. When we look directly at a line (*right*), the apparent motion is circular, producing circularly polarised photons.

motion it causes the particle to move in a circle. An electron in the stream will thus have a motion consisting of, firstly, its motion along the field line, and, secondly, a circular motion around the field line, and the two together produce a spiral.

Motion in a circle involves constant acceleration, and accelerating charges emit photons. Thus the spiralling motion results in radiation called *cyclotron emission*, which, for relatively slow-moving electrons, occurs at a characteristic *cyclotron frequency*.§ For faster motion the radiation occurs at integer multiples of the cyclotron frequency, refered to as cyclotron harmonics, and is smeared out into broad 'humps' around the frequency of each harmonic.

The spectra of AM Her stars thus contain 'cyclotron humps', as in Fig. 8.7. Measuring the wavelength of the humps, and deducing which harmonic creates each hump, yields the field strength in the region where the emission came from. This is one of the primary methods of measuring field strengths in these stars.

8.5.1 Polarisation

Cyclotron emission, in contrast to most radiation, is polarised; that is, the direction of the electric field it contains is not random. In normal light, the oscillating electric and magnetic fields that compose a light photon can point in any direction that is

---

§ The cyclotron frequency, ω_c, of an electron in a magnetic field B is eB/m_e rad s^{-1}, where e is the electron charge and m_e its mass. Thus the frequency of the emitted light is $\nu_c = \omega_c/2\pi = eB/2\pi m_e$ Hz.

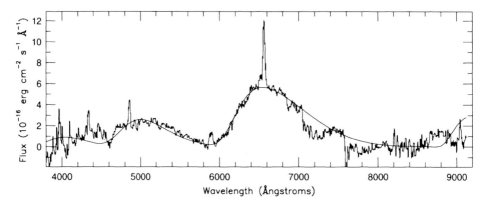

Fig. 8.7: A spectrum of the AM Her star RX 1313–32 showing cyclotron humps. and Balmer emission lines. The spectrum was taken during a low state (no accretion) and was processed by subtracting a blackbody spectrum (to remove the emission from the white dwarf) and the spectrum of a red dwarf. The humps are far less prominent when diluted by the accretion light of a high state. The fitted line is a model of the cyclotron emission, using a field strength of 56 MG. (Data and model by Hans-Christoph Thomas and colleagues.[7])

perpendicular to the direction of travel. However, if the electric vectors of a set of photons always point in one direction, the radiation is said to be *linearly polarised*.

Consider looking at a field line side on. From this viewpoint, an electron (which is spiralling around the field line) will appear to be oscillating perpendicularly to the field line. The photons produced by this motion will always have an electric vector in the direction of this oscillation, and so the light is linearly polarised.

Now consider viewing the field line head on. The electron will appear to be circling, and so the electric vector of the emitted photons (which follows the electron's motion) will be continually rotating, tracing out a circle. Such light is said to be *circularly polarised*.

Thus movement along field lines beams linearly polarised light perpendicular to the field line and circularly polarised light along the field line. Since the cyclotron emission can be as much as half the total light of an AM Her star, they are the most polarised objects in the sky. Indeed their nature was first recognised after the detection, by Santiago Tapia in 1976, of high levels of polarisation in the type star AM Her.[8] In recognition of this they are often called 'polars' as a synonym for 'AM Her stars'.

Polarisation is immensely important to the study of AM Her stars: from the ratio of linearly to circularly polarised light one can deduce the angle between the line-of-sight and the magnetic axis, and how it varies with orbital phase, thus gaining knowledge of the binary inclination and the tilt between the magnetic and spin axes. Modelling the polarisation can also yield the strength of the field; thus the geometries of AM Hers are often the best determined amongst cataclysmic variables.

Before leaving the effect of a magnetic field on light we should mention Zeeman

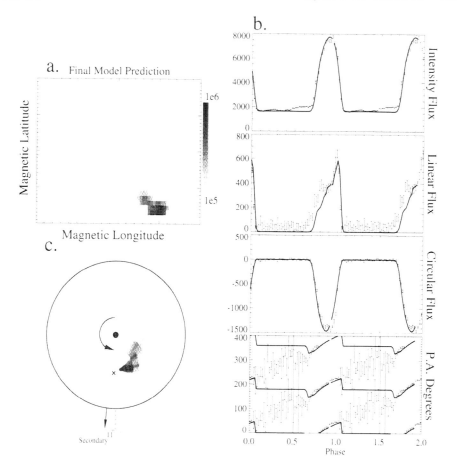

Fig. 8.8: Polarised light from ST LMi. Part B, top panel, shows the brightness over the orbital cycle, with a bright phase when the accretion region appears over the white-dwarf limb (cf. Fig. 8.5). Also shown are the amount of linearly polarised light, the circularly polarised light, and the position angle of the linearly polarised light (this is unreliable in the faint phase, where there is little polarised flux). Part A is a model of the accretion region, plotted in magnetic longitude and latitude. This is constructed by adding cyclotron emission from each pixel to obtain the best fit to the data (shown as solid lines in Part B). Part C shows the model accretion region plotted on the white-dwarf surface, in a view looking down on the 'south' pole of the white dwarf. The magnetic pole is marked with a cross. (Analysis and figures by Stephen Potter, using observations by Mark Cropper.[9])

splitting, the third major method for determining field strengths, along with polarisation and cyclotron harmonics. The orbits of electrons in atoms (set by quantum mechanics) are often oriented in a particular direction. Usually this does not affect the energy of the orbits, since one direction is as good as another. However, when

120 Magnetic cataclysmic variables I: AM Her stars   Ch 8

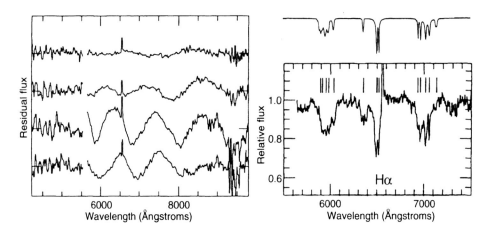

Fig. 8.9: *Left:* The cyclotron humps in MR Ser vary dramatically over orbital phase, since the viewing angle of the accretion region varies. The four spectra are from phases 0.41, 0.50, 0.58 and 0.67 from bottom to top, respectively. The data were taken in a low state and are shown after subtraction of a smooth continuum. The narrow line is Hα emission from the secondary. *Right:* Higher-resolution spectra reveal multiple Zeeman-split components of the Hα absorption from the white-dwarf surface. The individual components are marked by vertical lines, while a model Zeeman spectrum is shown above. Both the cyclotron humps and the degree of Zeeman splitting indicate a field strength of 28 MG. (Figures by Axel Schwope and colleagues.[10])

a magnetic field is applied, the energy of the orbit depends on its orientation with respect to the field. Thus the spectral lines resulting from transitions to this orbit are shifted slightly, split into several components depending on the possible orientations of the orbit. The degree of this *Zeeman splitting* depends on the magnetic field, and thus measuring the spacing of Zeeman-split lines in a spectrum yields the magnetic field strength.

8.6 THE ACCRETION REGION

It is worth investigating the details of the accretion region — the small portion of the white dwarf onto which the accretion stream falls — since it emits most of the radiation in an AM Her star.

The location of the accretion region is determined by the threading region; material threading onto a field line at a given radius from the white dwarf will follow that field line until it plunges into the white dwarf at a set distance from the magnetic pole.¶ However, since threading occurs at different locations depending

---

¶For threading at a radius R_{mag}, the accretion hits at a magnetic colatitude ϵ given by $\sin^2 \epsilon = R_{\mathrm{wd}}/R_{\mathrm{mag}}(1 - \sin^2\beta \cos^2\phi)$, where β is the angle between the magnetic and spin axes and ϕ the azimuthal angle away from the direction towards which the dipole tilts.

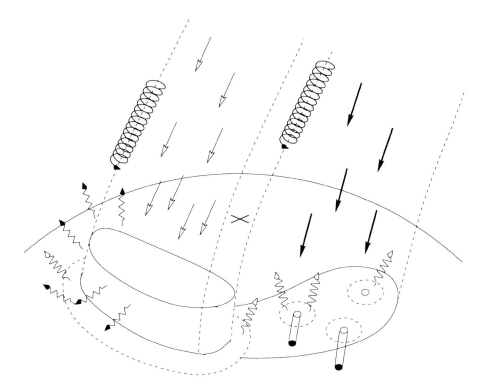

Fig. 8.10: A schematic AM Her accretion region. On the left diffuse material is funnelled by field lines into a ~ 20-keV accretion shock some distance from the magnetic pole, X. We see hard X-rays from the post-shock region and softer X-rays from the surrounding region, heated to ~ 20 eV by radiation from the shock. Electrons spiralling down the field lines also produce cyclotron emission. On the right, denser blobs of material plough deep into the white dwarf atmosphere, heating a region to soft-X-ray temperatures.

on the density of the material, we expect the accretion region to be stretched out into an arc near the magnetic pole, with denser blobs arriving preferentially at one end of the arc and more diffuse material arriving at the other end (Fig. 8.10).

The nature of the impact with the white dwarf depends crucially on the density of the accreting material, so we first consider the diffuse 'mist' and then the dense blobs.

As the diffuse mist hits the white dwarf its kinetic energy is converted to thermal energy, heating it to X-ray temperatures. This forms a pool of accreted material, which expands because it is hot. The result is a hot, dense *accretion column*, extending for $\sim 0.1\ R_{\rm wd}$ above the surface. The incoming material slams into the top of the column, where a shock forms, reducing the infall velocity by a factor 4. The liberated kinetic energy heats the material to $\sim 2 \times 10^8$ K (20 keV). At this temperature frequent collisions of electrons and ions result in the copious emission

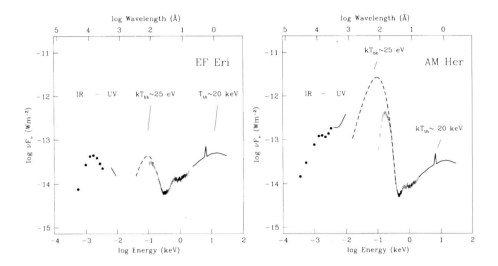

Fig. 8.11: Spectra of two AM Hers from the infrared to the hard X-ray, plotted as energy output versus photon energy. The three main components are (1) the hard bremsstrahlung emission from the accretion shock (which dominates at 1–100 keV); (2) soft-X-ray emission from blobby accretion and from reprocessing of harder X-rays (this dominates at 10–300 eV); and (3) the cyclotron emission in the optical and infrared. In EF Eri the three components emit similar amounts of energy, whereas in AM Her the soft-X-ray component is much larger, indicating that blobby accretion dominates the energetics. (Figures by Klaus Beuermann.[11])

of X-rays by the bremsstrahlung process; as it radiates the material slows further, cools, becomes denser, and so settles onto the white dwarf surface. The accretion columns are thus strong sources of hard-X-ray emission. Roughly half of these X-rays will be directed downward towards the white dwarf. Some will be reflected, but the majority will be absorbed and will heat the region around the column until it glows, emitting blackbody radiation at $\sim 200\,000$ K (20 eV).

Dense blobs are not affected by any accretion shock. They have sufficient momentum to plunge straight into the white dwarf, burying themselves deep in its atmosphere. Here their entire kinetic energy is absorbed by the white dwarf, and percolates to the surface, to emerge as more blackbody radiation at $\sim 200\,000$ K.[||] Thus in AM Her stars in which most of the accretion is in dense blobs, the dominant radiation is blackbody emission at soft-X-ray temperatures; indeed some AM Hers emit 50 times as much energy in soft X-rays as in hard X-rays (see Fig. 8.11).

A particularly revealing observation of AM Her itself was made by the *EXOSAT* X-ray satellite in 1983 (Fig. 8.12). The soft X-rays (upper panel) show the characteristic pattern of an accreting pole rotating from the visible face to the hidden face. Note also the intense, rapid variability of the X-rays when the pole is visible,

---

[||]The derivation of these temperatures is so similar to that in Box 7.1 that it need not be repeated.

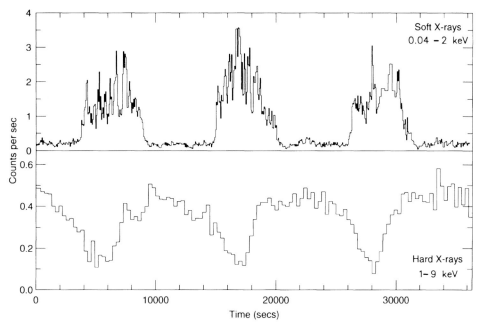

Fig. 8.12: A lightcurve of AM Her in soft X-rays (*top*) and hard X-rays (*bottom*). (Data from the *EXOSAT* satellite.[12])

which arises because individual blobs hit the surface in a random pattern, with ~15 accreting at any one time. The hard X-rays detected by *EXOSAT* (lower panel) are anti-phased with the soft X-rays, and so must come from the other pole, which is visible when the first pole is hidden, and vice versa. The explanation is that ~95% of the accretion was in the form of dense blobs flowing to one pole, to produce soft X-rays, while the remaining ~5% of diffuse material flowed to the other pole, created an accretion shock, and so produced hard X-rays.

8.7 ASYNCHRONOUS POLARS

Figure 8.13 shows optical lightcurves of V1432 Aql taken on two nights, a month apart. Both show a prominent eclipse (or possibly a dip caused by the accretion stream), recurring with a well-defined period of 12116 secs (3.37 hrs), but otherwise the modulation is different. An obvious broader dip seen near phase 0.7 on one night is not present on the other, and the remaining structure, though repeating from cycle to cycle on any night, is different on the two occasions. Although this star is an AM Her type, clearly something is changing on ~monthly timescales.[13]

Analysis of the photometry and of X-ray data reveals a second periodicity, at a slightly different period of 12150 secs. This suggests that the system is asynchronous: while the two stars orbit every 12116 secs, the white dwarf spins every 12150 secs, a difference of 0.3%. Because of the asynchronism the orientation of the white dwarf to the incoming stream changes slowly (unlike in normal AM Her stars), so that the secondary star appears to make one orbit every 50 days, as seen by someone rotating with the white dwarf. This 'beat' cycle for the relative orientation of the two again follows the formula from Box 6.1. As the orientation changes, the stream will first feed one magnetic pole, then flip to the other pole as it becomes more favourably presented, then flip back to the first pole to complete the cycle. The different accretion geometry at different parts of this beat cycle accounts for the difference in the lightcurves a month apart.

We currently know of four systems in which the white dwarf and the orbit are asynchronous by $\sim 1\%$ (V1432 Aql, V1500 Cyg, BY Cam and CD Ind), comprising 10% of the known AM Her stars. Although asynchronous rotation strictly violates the membership rule of the AM Her class, the beat cycle in such systems is so long that they behave as an AM Her over any one orbit, and so are conventionally classed with the AM Hers.

8.7.1 The origin of synchronous rotation

The question of why 10% of the AM Hers are asynchronous is best approached by asking why the rest are synchronised. The usual effect of accretion is to spin up the white dwarf, since the angular momentum of the accretion stream (originating from the orbital motion of the secondary) is deposited on the white dwarf. Thus the white dwarfs in many non-magnetic cataclysmics are thought to rotate with periods as short as ~ 50 secs.

Since white dwarfs in AM Hers rotate only every 1–3 hrs, something must be

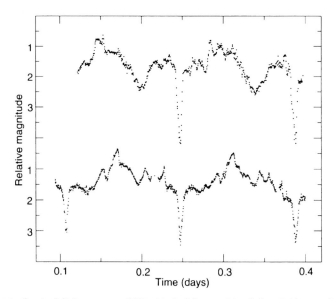

Fig. 8.13: Optical lightcurves of V1432 Aql from 1994 July 16 (*bottom*) and 1994 August 14 (*top*). (Data by Joe Patterson and colleagues.[13])

Sec 8.7   Asynchronous polars   125

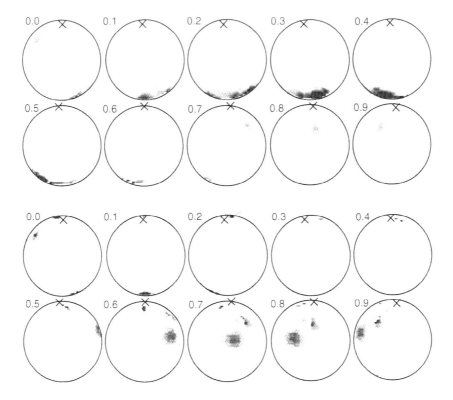

Fig. 8.14: The accretion regions of the asynchronous AM Her CD Ind, shown at ten different orbital phases as the binary rotates. The top set result from data at phase 0.55 of the 6.3-day beat cycle, while the lower set are from phase 0.15. This illustrates how the relative geometry and thus the accretion 'footprints' change as the magnetic dipole alters its orientation with respect to the secondary. The mapping uses the techniques outlined in the caption to Fig. 8.8, and are from work by Gavin Ramsay, Stephen Potter and colleagues.[14]

slowing them down. This is thought to be the interaction of the strong magnetic field of the white dwarf with the field of the secondary. The two fields can intertwine where they meet, entangling their field lines. Thus if one is spinning faster, the joint field lines will be 'wound up', increasing their tension and stored energy and creating a drag force. This drag acts as a torque on the white dwarf, counteracting the effect of accretion, and slowing it down into synchronous rotation.

Asynchronism can thus result if the sychronising torque is not quite strong enough to do its job, for instance if the white-dwarf field is weaker than average, or the binary separation is larger (recall that field strength declines with distance). The observed asynchronous systems may also be in short-term departure from synchronism. For instance, V1500 Cyg underwent a nova eruption in 1975 (see Chapter 11), which is thought to have knocked the system out of synchronism. Indeed, the white dwarf, currently spinning 1.7% faster than the orbit, is observed to be slowing down sufficiently rapidly that it will regain synchronism in ~ 170 years.[15]

Chapter 9

Magnetic cataclysmic variables II: intermediate polars

In the majority of cataclysmic variables the magnetic field is sufficiently weak that it can be ignored, whereas in AM Her stars it completely dominates the accretion flow. It is the intermediate case, however, that leads to the greatest potential for complexity. With a medium-strength field a cataclysmic variable can combine the characteristics of a non-magnetic system (in its outer regions) with those of an AM Her (nearer the white dwarf) and produce new phenomena where they interact.

We have already started along this road at the end of the last Chapter with a discussion of asynchronous polars, showing that AM Hers can be knocked out of synchronism if the field is a little too weak or the stellar separation a little too large. Continuing this trend leads to systems which have lost synchronism entirely, in which the white dwarf is spun up by the accretion of material, ending at rotation periods of typically a tenth of the orbital cycle. Such systems are called 'DQ Her stars' or *intermediate polars*,[1] to denote a status half way between AM Her stars (polars) and the non-magnetic cataclysmic variables.*

9.1 V2400 OPH: A DISCLESS INTERMEDIATE POLAR

Consider the lightcurves of V2400 Oph, which are presented in the form of *Fourier transforms* in Fig. 9.1. Fourier transforms display the different frequencies that constitute a lightcurve (see Box 9.1), in the same way that a sample of music can be decomposed into frequencies by a 'frequency analyser'.

The uppermost Fourier transform from V2400 Oph reveals that the circularly polarised light varies with a period of 927 secs. Since the polarisation is caused by the field, this must be the period at which the magnetic dipole, and thus the

---
*Historically, intermediate polars (with typical spin periods of 1000 secs) have been distinguished from DQ Her stars (with faster spin periods of typically 50 secs) and are thus 'intermediate' between DQ Hers and AM Hers. However, since the distribution of spin periods is a continuum rather than a dichotomy, the terms 'intermediate polar' and 'DQ Her star' are increasingly regarded as synonymous (cf. 'polar' and 'AM Her star'), denoting all magnetic systems that are not in strict nor approximate synchronism. Other authors may retain the distinction.

128   Magnetic cataclysmic variables II: intermediate polars   Ch 9

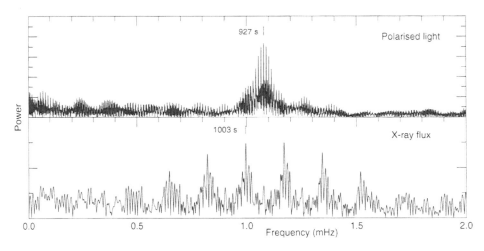

Fig. 9.1: The Fourier transform of the polarised light of V2400 Oph reveals the 927-sec spin period of the white dwarf. The X-ray flux, however, varies at the 1003-sec beat period, implying that V2400 Oph is a discless accretor. Both transforms also contain 'alias' peaks of the true periodicities. (Data by David Buckley.[2])

white dwarf, is spinning. The X-ray flux, however, varies with a period of 1003 secs, while spectroscopy of the star shows that the orbital period is 3.4 hrs. Since 1003 secs is the beat period between the spin and orbital cycles, the explanation is straightforward: the accretion stream is flipping between the two magnetic poles as the white dwarf spins beneath it, and thus produces an X-ray flux that varies with the interaction period.[†] This is confirmed by Doppler shifts of the emission lines (Fig. 9.2) which vary over the beat period, indicating that the infalling material periodically changes direction to flow to the other pole.

9.1.1 The spin period of a discless accretor

Can we explain why the spin period of V2400 Oph is about 8% of the orbital period? The stream of material will emerge with an angular momentum appropriate to the Lagrangian point, which is equivalent to that in a Keplerian orbit at the circularisation radius (see Section 2.4). This means that the white dwarf will tend to adjust its spin rate so that, at this circularisation radius, the magnetic field lines are travelling at the same speed as the local Keplerian orbit.

To understand this, consider what happens if the magnetosphere is spinning more slowly. The accreting material will then have a greater angular momentum (per unit mass) than the white dwarf, and so will spin up the white dwarf. If the magnetosphere were spinning more rapidly, however, the stream material could not attach to field lines at the circularisation radius, since it would have to gain angular momentum to do so. There are then two possibilities. If the field at the

---

[†]With ω and Ω as the spin and orbital frequencies, respectively, the beat frequency is $\omega - \Omega$. Equivalently, $1/P_{\text{beat}} = 1/P_{\text{spin}} - 1/P_{\text{orb}}$; cf. Box 6.1.

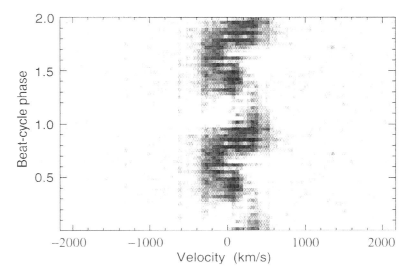

Fig. 9.2: The Hβ line of V2400 Oph shifts from blue to red over the beat cycle, as the stream flips from pole to pole.

circularisation radius were strong enough to control the accretion flow, the rapidly spinning magnetosphere would form a 'centrifugal barrier' preventing accretion, and would expel the material (as a child would be flung from a roundabout that was spinning too rapidly). If the field were too weak to dominate at the circularisation radius, the stream's angular momentum would keep it in a circular orbit at this radius, and it could accrete only by spreading into a disc, thus pushing some material inwards to connect onto the field lines.

The upshot of the above is that a discless accretor has an equilibrium situation, occurring when the circularisation radius is the same as the *corotation radius* (the point at which the magnetic field corotates with the local Keplerian orbit).[‡] Given formulae for both radii (see Boxes 2.1 and 2.4) we find that $P_{\rm spin} \approx 0.07 P_{\rm orbit}$ for reasonable values of the stellar masses, in agreement with the observed value for V2400 Oph.[3]

If a disc forms, the material will spread inward from the circularisation radius, onto faster orbits with a lower angular momentum. The resulting equilibrium spin period would be less than the above value. Indeed, on a plot of $P_{\rm spin}$ versus $P_{\rm orb}$ (Fig. 9.3), all the known systems are on or below a line of $P_{\rm spin} = 0.1 P_{\rm orb}$ with the exception of EX Hya and V1025 Cen. These two are either far from equilibrium (perhaps through a change in mass-transfer rate) or have found a different equilibrium. One suggestion is that they possess stronger fields that resonate with the Lagrangian point, rather than with the circularisation radius.[4]

[‡] Equating $v_{\rm Kep} = [GM_{\rm wd}/r]^{1/2}$ with $v_{\rm mag} = 2\pi r/P_{\rm spin}$ gives $r_{\rm co\text{-}rot} = [P_{\rm spin}^2 GM_{\rm wd}/4\pi^2]^{1/3}$.

Box 9.1: Fourier analysis

Full mathematical accounts of Fourier analysis can be found in many mathematical textbooks, but rather than present another here, I describe how the basic principles can be understood non-mathematically.

The fundamental starting point is that any lightcurve (indeed any record of how some quantity changes with time) can be reproduced by summing a set of sine waves. To reproduce a continuous, infinitely long lightcurve would require an infinite number of sine waves with frequencies ranging from zero to infinity. However, in practise a finite set of sine waves suffices, since, firstly, periods much longer than the data train are not relevant. Secondly, a lightcurve is limited by how often it is sampled, and periodicities faster than this (strictly, frequencies faster than the 'Nyquist frequency' of $f = 1/2\Delta$ where Δ is the sampling time interval) are not recoverable, though they can produce spurious 'aliases' in the transform.

The amplitudes of each sine wave then reveals how much of each frequency is present in the lightcurves, and are plotted against frequency to produce a Fourier transform (alternatively it is common to plot the power at each frequency, which corresponds to the square of the amplitude; the other information, the phase of each sine wave, is usually not plotted).

Several complications often arise. First, if a pulsation has the shape of a sine wave, it will produce a single peak in the transform. A more complex pulse shape, however, is reproduced by adding in sine waves at twice, three times, etc, the pulsation frequency. These are called *harmonics* of the fundamental frequency, and are often present in transforms.

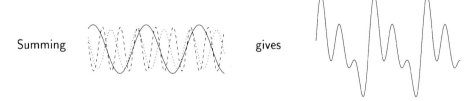

Complex waveforms result from adding harmonics to a fundamental sinusoid.

Secondly, the accuracy to which a peak can be located in frequency depends on the length of the data train. A long pulse train allows an accurate measure and thus produces a narrow peak in the transform, whereas a shorter train produces a broader peak. In fact, the frequency resolution equals the reciprocal of the data length.

Longer data trains correspond to increased frequency resolution.

Thirdly, a lightcurve is often unevenly sampled; for example it can contain gaps due to clouds or daylight. This leads to the issue of *aliasing*. The amount of data either side of a gap might only determine a pulsation frequency to an accuracy which is insufficient to count, with certainty, the number of pulse cycles across the gap. This ambiguity produces a set of alias peaks, with separations in frequency corresponding to one cycle more or fewer across the gap. With complex sampling patterns, very complex aliasing structure can result. As a rule of thumb, the highest alias peak is almost always the true period if it is $\gtrsim 30\%$ higher than the next highest alias; if this is not the case there may be no way of resolving the ambiguity without more data.

When the data (bold line) contains a gap, several alias frequencies (differing by integral cycles across the gap) can fit the data acceptably. In the resulting Fourier transform (right) the width of the individual peaks is set by the length of the entire train, while the width of the envelope of alias peaks is set by the length of the longest uninterrupted data section.

A last, and more difficult, issue is whether a peak in a transform is high enough to indicate a true periodicity in the system, or whether it results from a chance fluctuation. In cataclysmic variables there is no clear way of determining this, since the level of random fluctuations (or 'flickering', see Chapter 10) is large, variable, and unknown. Without an accurate model of the flickering no formal statistical test will work. The starting point is to compare the suspect peak with the 'noise' level due to flickering at surrounding frequencies, but beware that of course all transforms contain a highest peak. The only reliable test is repetition — looking for a suspect periodicity in several datasets, or dividing a dataset into sections to determine whether it is present in each — since by definition flickering will not recur preferentially at one frequency. The literature on cataclysmic variables contains many false claims of periodicities where such tests and appropriate scepticism have not been applied.

132 Magnetic cataclysmic variables II: intermediate polars   Ch 9

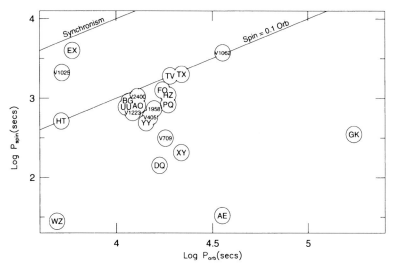

Fig. 9.3: Twenty-one known intermediate polars located on a plot of spin period versus orbital period. The stars are identified by abbreviated names.[5]

9.1.2 Accretion of diamagnetic blobs

Andrew King and Graham Wynn have extended the analysis of a discless intermediate polar,[6] following the realisation that accretion streams can break up into blobs of different densities (as shown by observations of AM Her stars; Section 8.6). Each blob can then be regarded as interacting independently with the spinning magnetic field. The field cannot penetrate the blobs but induces electric currents on their surface (termed 'diamagnetic' behaviour). If the blobs and the field lines are moving at different rates, the surface currents produce a 'drag' force proportional to the difference in velocity, changing their motion from a pure free-fall.

The fate of individual blobs depends on their energy. Blobs with lower-than-average energy will spiral inwards, losing energy as they drag against the field, and be accreted onto the white dwarf. Blobs of average energy will tend to congregate at the circularisation radius, orbiting many times before they dissipate enough energy to start spiralling inwards. Blobs of greater energy can spiral outwards, gaining energy in their interaction with the field (for radii larger than the corotation radius the field-line motion is faster than a Keplerian orbit, so the blobs are sped up by the drag force). Such blobs are eventually swept up by the secondary star. Thus discless accretion can lead to a torus of blobs orbiting at the circularisation radius, with a smaller number spiralling both inwards and outwards.

9.2 DISC FORMATION

Under what circumstances does the interaction of a stream with a magnetic field give way to the formation of a disc? An easy case to deal with is when the magnetosphere

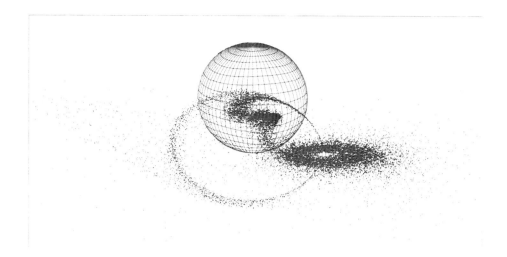

Fig. 9.4: A snapshot from a simulation of a discless intermediate polar, by Graham Wynn. The stream of material from the secondary star, modelled as diamagnetic 'blobs', encounters the magnetic field. Depending on their energy and density, the blobs are trapped in magnetic loops, expelled back towards the secondary, or can spiral inwards to form a disc-like structure and so accrete onto the white dwarf.[4]

is smaller than the radius of minimum approach of the free-falling accretion stream ($r_{\text{mag}} < r_{\text{min}}$; recall Fig. 2.7). The stream could then circle the white dwarf and form a disc, ignoring the feeble magnetic field. The disc would then spread (inwards and outwards) until its inner edge encountered the magnetosphere.

An opposite case is when the magnetosphere is significantly larger than the circularisation radius ($r_{\text{mag}} > r_{\text{circ}}$). Any attempt to form a disc will now fail, since the average angular momentum of the stream is insufficient to allow it to orbit at radii greater than the circularisation radius. The accretion would remain stream-fed, as in V2400 Oph.

The intermediate case ($r_{\text{circ}} > r_{\text{mag}} > r_{\text{min}}$) is less clear. As discussed above, dense diamagnetic blobs may succeed in making several orbits, despite the field. If so, currents induced on their surfaces would oppose the magnetic field, screening it from blobs further out, and helping them to ignore the field. This could lead to a build-up of blobs, feeding back into increased screening, and thus allowing sufficient material to accumulate near the circularisation radius to form a disc.

9.2.1 Diagnosing the accretion mode

Evidence for the presence of discs (or at least of circling material that has spread out into a uniform ring, thus mimicking a disc) comes from the X-ray lightcurves of intermediate polars. When the disc is uniform in orbital phase, the feeding of accretion onto field lines varies only with the phase of the white-dwarf spin cycle.

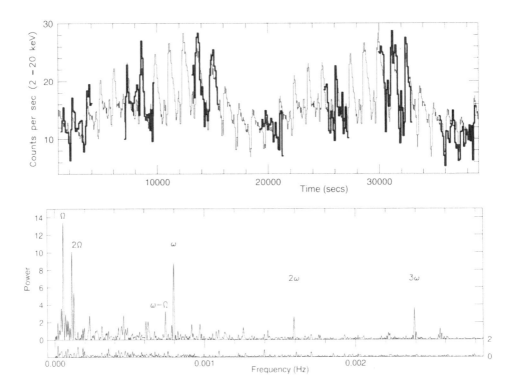

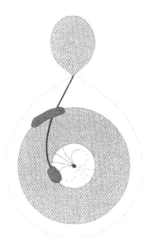

Fig. 9.5: An X-ray lightcurve covering two orbital cycles of FO Aqr (*top*), from the *Ginga* satellite. The lighter line is a model based on the data. The data themselves (heavier line) are more extensive than shown but are too broken up for easy display. The Fourier transform of the data (*middle*) reveals pulses at the 1254-sec spin period (frequency ω and its harmonics 2ω and 3ω). A pulsation at the 1351-sec beat cycle (frequency $\omega - \Omega$, where Ω is the orbital frequnecy) interferes with the spin pulses so that the pulse profile changes with orbital phase. There is also a strong 4.85-hr orbital modulation (frequency Ω and its harmonic 2Ω). The remaining peaks are aliases, while the lower panel is the transform after subtracting the spin, beat and orbital pulsations.

The beat pulse results from disc-overflow accretion, as illustrated *left*. The orbital modulation is 'dipping' behaviour, seen at phases when the overflowing stream obscures the white dwarf (note that FO Aqr has a high-enough inclination to show a grazing eclipse). All three modulations are different in different observations (sometimes the beat pulse is absent) indicating that disc-overflow accretion is highly variable.[7]

Thus the emitted X-rays are pulsed solely at the spin period. Accretion that remains localised in orbital phase, such as the stream accretion in V2400 Oph, leads instead to X-rays pulsed at the spin–orbit beat frequency, since this is the frequency at which the accretion interacts with the field.[8]

It is notable that in most intermediate polars the X-ray spin pulses are always larger than the X-ray beat pulses (the exceptions are V2400 Oph and TX Col). This implies that the bulk of their accretion flow passes through a disc. Furthermore, the fact that one often sees 'bright spots' where the stream collides with the disc, and which imply a disc radius much greater than the circularisation radius, suggests that these are true discs, not just rings of blobs.

However, X-ray beat pulses are seen in many systems, so some fraction of stream-fed accretion does occur, often concurrently with disc-fed accretion (see Fig. 9.5). This is reminiscent of many non-magnetic systems, notably SW Sex stars (Section 7.4), where there is evidence that the accretion stream overflows the initial impact with the disc and continues onward to near the white dwarf. In intermediate polars the overflowing stream would encounter the magnetosphere and so give rise to the X-ray beat pulses.[§] Furthermore, such 'stream/disc overflow' appears to be highly variable, judging from the large changes in the relative amplitudes of the spin and beat pulses, as measured in X-ray lightcurves from different observations.[9]

9.3 DISC-FED ACCRETION IN AN INTERMEDIATE POLAR

Whereas AM Her stars are certified by the observation of polarised light — a direct indicator of a magnetic field — this is present in only 3 out of 24 known intermediate polars. The lack of polarisation results from a lower field strength (typically 1–10 MG in intermediate polars compared to 10–100 MG in AM Hers) and from the dilution of the polarised light by bright discs. Instead, we infer the presence of a magnetic field from the observation of pulsed X-rays, the hallmark of an intermediate polar. The dominance of spin-cycle pulses over beat-cycle pulses then leads to the 'standard' picture of an intermediate polar, involving a disc which, at large radii, is little different from that in a non-magnetic system. Only inside the magnetosphere is the disc disrupted, and replaced by inflow along field lines.

In equilibrium, a disc-fed white dwarf will corotate with the Keplerian motion at the magnetosphere.[¶] If the spin were faster, the field lines would drag through the material at the inner edge of the disc, resulting in a braking torque on the white dwarf; if the spin were slower, the opposite effect would spin up the white dwarf. The opposing torques thus keep the spin period near equilibrium, despite the accretion of material. In FO Aqr we see the spin period wander around what we presume is the equilibrium value (see Fig. 9.7), changing from spinning down to spinning up on a ∼ 10-yr timescale. (Most other systems show only one or the other,

---

[§]Another possibility is that spiral shocks in a disc might also lead to beat periods, since the shocks are fixed in orbital phase.[10] However, since shocks are two-armed spirals, they lead naturally to pulses at twice the beat frequency, rather than the beat frequency itself.

[¶]That is, $r_{\text{co-rot}} = r_{\text{mag}}$, and both are less than r_{circ}.

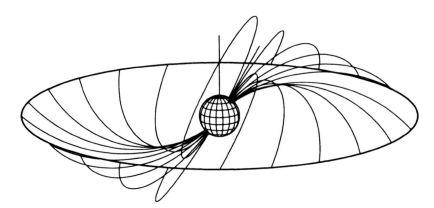

Fig. 9.6: The pattern of field lines feeding from the inner edge of a disc (outer ring) onto the white dwarf. One side of the disc feeds the upper pole while the other side feeds the lower pole. Straight lines from the white dwarf mark the spin axis and the offset axis of the magnetic dipole. (Figure by Andrew Beardmore.)

but this is presumably because our observation spans are very short compared to the timescale on which the white dwarf adjusts its spin period.)

The field–disc interaction, though, is further complicated by the tilt of the magnetic field with respect to the rotation axis of the white dwarf (just as magnetic north on Earth is offset by 12° from the North Pole). Thus the field strength at any one radius varies over the spin cycle (if this were not the case, nothing would change over the spin cycle, and no pulsation would be seen).

It then makes sense to suppose that the accretion flowing to the upper pole is picked up from the region of disc to which it points, with accretion from the opposite side of the disc flowing to the lower pole (see Fig. 9.6). This is energetically favourable, since otherwise the accretion would have to go the long way round the

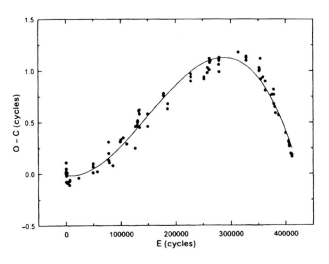

Fig. 9.7: The '$O - C$', or observed minus calculated diagram for the white-dwarf spin cycle of FO Aqr. It shows the white dwarf spinning down to a slightly longer period, and then spinning up again. (Figure by Joe Patterson.[11])

field lines, diverting further out of the plane.

Accretion onto two poles creates a fundamental difference between the lightcurves of intermediate polars and those of AM Hers. The single accretion region in AM Hers can be on the far side of the white dwarf for roughly half the spin cycle, producing episodes of zero X-ray flux (e.g. Fig. 8.5). In intermediate polars the accretion regions near the two poles are diametrically opposite on the white dwarf: when one disappears over the white-dwarf limb the other will simultaneously appear. Thus one of the two regions is always in view, and the X-ray flux never reaches zero.

9.3.1 The accretion footprints

Tracing back field lines from the disc-disruption radius to the white-dwarf surface leads to an accretion 'footprint' in the shape of a ring around the magnetic pole (this is equivalent to the ring of aurora circling Earth's magnetic poles). If accretion feeds from only a range of azimuth, however, the footprint is an arc-shaped segment of the ring. The angle between the accretion ring and the magnetic pole is set by the disc-disruption radius, in that a smaller disruption radius feeds field lines that enter the white dwarf further from the poles (see footnote on page 120).

To investigate further we need to ask such questions as: over what range of azimuth does disc material feed onto field lines? (This sets the length of the accretion arc.) Over what range of disc radii does material feed onto field lines? (This sets the width of the accretion arc.) Are the field lines distorted and twisted by the interaction with the disc (which would change the accretion footprint)? Is the inner disc warped by the presence of the field, particularly since the local field strength varies periodically with spin phase? There is no settled answer to any of these questions, which are at the limit of current theory.[12] We can, though, consider some observational clues, and in particular the instructive case of XY Ari.

9.3.2 Accretion in XY Ari

XY Ari is seen nearly edge on, creating deep X-ray eclipses every 6.06-hr orbital cycle (unfortunately it lies behind a molecular cloud, and so is invisible in the optical). Fig. 9.8 shows it emerging from eclipse and pulsing at the 206-sec spin period. An analysis of twenty such eclipses showed that the bulk of the X-ray flux emerges from eclipse in less than 2 secs, whereas the white dwarf takes ≈ 30 secs to emerge. This implies that most of the X-rays originate from a small, concentrated accretion region covering only ~ 0.001 of the white dwarf surface (although there is evidence for an additional accretion region which is more extended but fainter).[13] Such a small area implies that both the radial and azimuthal extents of the threading region are small.$^{\|}$

Furthermore, timing the eclipse egresses at different phases of the spin cycle reveals the location of the accretion region on the white dwarf. For half the spin

---

$^{\|}$For instance, a radial extent of $0.1\,r_{\rm mag}$ and an azimuthal extent of $30°$ yield an accretion area, f, of ~ 0.001, as a fraction of the white dwarf surface; however the observational uncertainties are such that these numbers are only illustrative.

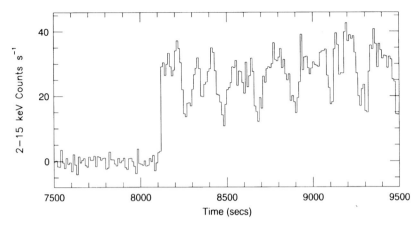

Fig. 9.8: The X-ray lightcurve of XY Ari as it emerges from eclipse. The data, from the *RXTE* satellite, were obtained during outburst when the 206-sec pulsation had a higher amplitude than in quiescence.[13]

cycle the upper pole tracks across the visible face, then for the remaining half the lower pole is seen. The result is a near-constant X-ray flux. The low-level pulsation normally seen in XY Ari must result from a slight asymmetry between the poles, such that the appearances and disappearances of the two poles are not quite synchronised.

9.3.3 XY Ari's response to a disc instability

In addition to being an intermediate polar, XY Ari is a dwarf nova, meaning that outside its magnetosphere resides a disc capable of undergoing the dwarf-nova instability. Fortuitously, an outburst was monitored by the *RXTE* X-ray satellite, and the resulting data illustrate how a magnetosphere responds to a changing accretion rate.[14]

The change in the emitted X-ray flux over the outburst is shown in Fig. 9.9, together with the X-ray pulsation profile at three stages in the outburst. At the start of the outburst the amplitude of the pulsation jumps dramatically. The reason for this relates to XY Ari's very high inclination of $\approx 82°$. With the quiescent magnetosphere extending to 9 white-dwarf radii (estimated by assuming corotation with the 206-sec spin period), we can only just see the lower regions of the white dwarf above the inner edge of the disc (see Fig. 9.9). The extra viscosity of a disc instability, however, pushes a flood of extra material towards the magnetosphere. This boosts the disc's ram pressure, allowing it to overwhelm the magnetic field and push the magnetosphere inwards (see Box 8.1). For a factor 24 increase in the accretion rate (judged by the increase in X-ray flux during the outburst) the disc pushes inwards to ≈ 4 white-dwarf radii. This means that our view of the lower pole of the white dwarf is cut off. The symmetry of the quiescent situation is then broken; the disappearance of the upper pole over the white dwarf limb is no longer compensated for by the appearance of the lower pole, and so a much stronger

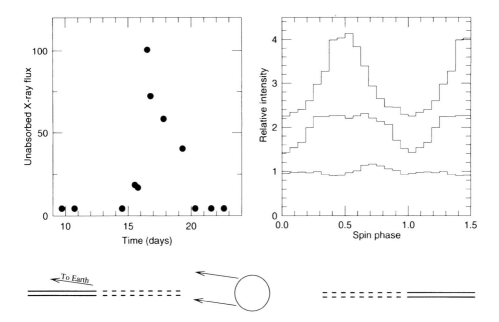

Fig. 9.9: *Left:* the X-ray flux of XY Ari over an outburst. *Right:* the change in the X-ray pulse profile from quiescence (lowest curve) to the first day of outburst (middle) and the peak of outburst (upper); the zero-points of the three curves are at 0, 1 and 2 respectively. *Lower:* An edge-on scale drawing of the inner disc of XY Ari. In quiescence (solid line) the hole in the disc allows us to see the whole white dwarf; in outburst (dashed line) the disc pushes inwards and cuts off the view of the lower pole, thus dramatically changing the X-ray pulse profile.[14]

'AM Her-like' modulation is seen.

Thus, on the first day of outburst, the pulse amplitude jumped as the disc pushed inwards, but the X-ray flux increased only slightly. Presumably, most of the extra material went to filling the new region of disc, rather than accreting. Only when the disc reached the new equilibrium, appropriate to the increased mass flow, did the X-ray flux climb dramatically.

The details of the pulse profiles are also interesting. On the first day of outburst the profile was flat-topped, implying that at some spin phases the entire accretion region lay on the visible face. However, by the second day the profile was continually varying, implying that some regions of the pole were continually appearing or disappearing over the white-dwarf limb. Thus the accretion regions must have grown much larger (Fig. 9.10). One factor is that the decrease in the magnetospheric radius maps to a larger ring around the magnetic pole. (An equivalent situation occurs when our Sun emits a burst of dense solar wind in a 'coronal mass ejection'; this crushes Earth's magnetosphere and the ring of aurora expands. Aurorae are then seen at much lower latitudes than normally.) A second factor is that

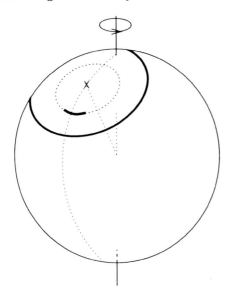

Fig. 9.10: The accretion region on XY Ari's white dwarf. In quiescence, accretion falls onto a relatively short arc of a ring around the magnetic pole (X), implying that accretion feeds only from the region of inner disc at which the magnetic axis points. In outburst, the accretion arrives further from the magnetic pole, since the magnetosphere is smaller. The accretion ring is now nearly complete, implying that disc material feeds onto field lines from nearly all directions.

accretion was feeding over a much greater range of azimuth then in quiescence, increasing from a short arc of $\sim 30°$ to a nearly complete ring covering $\gtrsim 210°$. This probably occurred because the system was far from equilibrium with the shrunken magnetosphere: the Keplerian motion at the inner disc would have exceeded the corotation velocity, and the extra kinetic energy of the accreting material allowed it to take energetically less favourable paths to the white dwarf that are forbidden in quiescence.

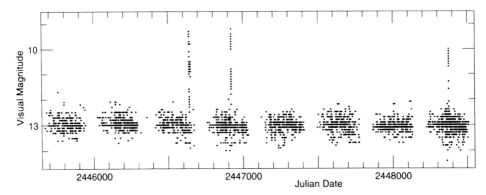

Fig. 9.11: EX Hya is an intermediate polar with an orbital period of only 98 minutes, and so is expected to be a dwarf nova. However, its outbursts are peculiar, being rare, erratic, and lasting for only 2–3 days. It is unclear whether this difference results from the magnetic disruption of the inner disc, or whether EX Hya's outbursts are a different phenomenon. The plot shows four outbursts in eight years: the first two, in 1986, were only eight days apart, and are barely resolved on this scale. The third and fourth occurred in 1987 and 1991. (Data compliation by the Variable Star Section of the RASNZ.[15])

9.4 PULSATIONS IN A DISC-FED ACCRETOR

If disc-fed accretion in intermediate polars lands at both magnetic poles, and if the occultations of one pole are cancelled out by the other pole, why are strong X-ray pulsations so characteristic of intermediate polars? (Note that most have a much lower inclination than XY Ari, so blocking one pole with the disc is not feasible.) One possibility invokes asymmetries between the two poles: if the magnetic dipole is slightly off-centre in the white dwarf, occultation effects will not exactly cancel. Also, if the X-ray-emitting accretion regions have a significant height above the white-dwarf surface, this also breaks the symmetry: a height of 0.1 white-dwarf radii allows X-rays to be seen for 0.07 of a cycle longer as a region passes over the limb. Although such effects undoubtably occur, they lead only to low-level pulsations, such as those in the quiescent XY Ari. They cannot explain the much stronger pulsations typified by those of AO Psc (Fig. 9.12).

The biggest clue is that X-ray pulsations in intermediate polars are deeper at lower energies. This is characteristic of absorption of X-rays by material in the accretion flow, since lower-energy X-rays are absorbed more effectively.

Thus the minimum of the pulsation occurs when the upper magnetic pole points towards us; our view of the upper accretion region is then obscured by the infalling material, and much of the X-ray flux is absorbed or scattered out of the line of sight.[16] The lower pole is at that point hidden behind the white dwarf, and so contributes no flux.

Half a spin-cycle later, when the upper pole points away from us, our view of the accretion regions is unobscured and we see more X-rays (note that this applies to both poles, and one of these will always be in view; see Fig. 9.13)

This picture is refered to as the *accretion curtain* model, in which the pulsation is attributed primarily to the veiling effect of the 'curtains' of magnetically channeled accretion flow. One of the main pieces of supporting evidence is that the optical line emission is blue-shifted at X-ray pulse maximum. Since the line emission comes preferentially from the most visible pole (presumably the upper one) it implies that material falling to that pole is travelling towards us at X-ray maximum, in agreement with the hypothesis that the upper pole is then pointing away from us.[17]

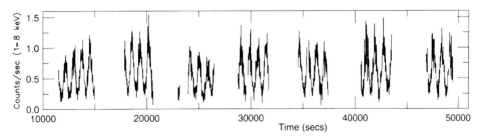

Fig. 9.12: An X-ray lightcurve of AO Psc, showing deep 805-sec pulsations. The data, from the *ASCA* satellite, are broken by the satellite's low Earth orbit.[18]

142   Magnetic cataclysmic variables II: intermediate polars   Ch 9

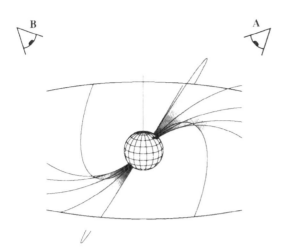

Fig. 9.13: The 'accretion curtain' model suggests that we see maximum light when the optically thick curtains are viewed side on (from B), projecting most surface area. Pulse minimum occurs half a spin-cycle later (viewing from A) when flux emitted in our direction is absorbed by the curtain itself, and the lower pole may be hidden by the white dwarf. (Illustration by Andrew Beardmore.)

9.4.1 Optical pulsations at the spin period

The accretion curtain model can also explain the fact that spin-cycle pulsations are often seen in the optical lightcurves of intermediate polars. The curtains themselves will be optically bright, and since they are azimuthally extended, maximum curtain area will be on display when the upper pole points away from us, producing a maximum in the optical light in phase with the X-ray maximum.

The strongest evidence for this idea comes from optical eclipses of EX Hya, which has a grazing eclipse. In fact the eclipse line in EX Hya passes through the white dwarf, so that the lower accretion curtain and the lower pole are eclipsed, but the upper curtain remains uneclipsed (we know this because in mid-eclipse the X-rays are reduced by roughly half, implying that one of two poles is occulted).

The first point is that the optical eclipse depth is heavily correlated with the phase of the optical spin pulse, implying that the region responsible for the pulsation is being eclipsed. Secondly, the optical eclipse is slightly early at one phase

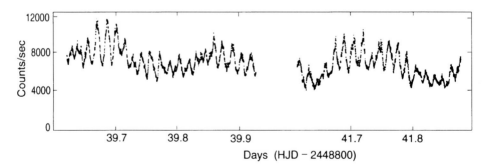

Fig. 9.14: The optical lightcurve of FO Aqr shows a large-amplitude pulsation at the 1254-sec spin period and a variation at the 4.85-hr orbital period. Fourier analysis also reveals weaker sidebands (see Box 9.2). (Data by Joe Patterson.[11])

Sec 9.4 Pulsations in a disc-fed accretor 143

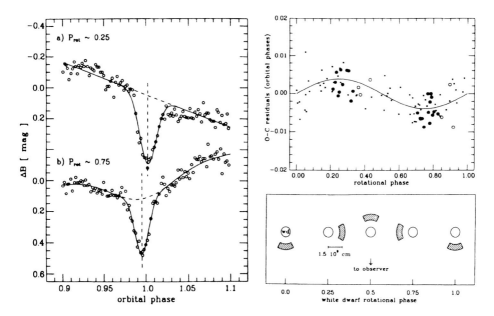

Fig. 9.15: The time of eclipse in EX Hya's optical light varies slightly with the phase of the spin cycle (*top right*), implying that the eclipsed emission originates in the 'accretion curtains' of material falling onto the magnetic poles. In the grazing eclipse, only the lower accretion curtain is occulted, and this is eclipsed slightly earlier at spin phase 0.75, and slightly later at spin phase 0.25 (*left*). From the size of the effect one can deduce that the centroid of the optical emission from the lower curtain is ≈ 2 white-dwarf radii from the white-dwarf surface (*bottom right*). (All figures by Siegel and colleagues.[19])

of the spin cycle, and slightly late at the opposite spin-phase (see Fig. 9.15). This implies that the region being eclipsed corotates with the white dwarf. Furthermore, the amplitude and phase of the wandering of the eclipse locate the region at two white-dwarf radii from the white dwarf, phased so that it is closest to us at X-ray-and-optical pulse maximum. This is just as predicted for the lower accretion curtain.**

9.4.2 Double-peaked pulsations

The accretion curtain model adequately explains the roughly sinusoidal pulsations seen in many intermediate polars. However, some systems instead show a pulsation with two peaks per cycle. How can we explain this? The two peaks suggest the involvement of two poles, but we know that two poles don't necessarily lead to

---

**Intermediate polars are never straightforward, and even though the optical pulse in EX Hya is explained well by the accretion curtain model, it may not explain the X-ray pulse. Instead, atypically tall accretion columns, resulting from an atypically small accretion rate, may disappear only partially over the limb, thus producing the observed pulsation.

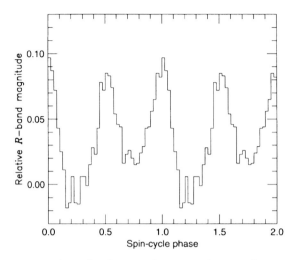

Fig. 9.16: The spin period of V405 Aur was originally thought to be 272 secs, but from noticing that every second trough was deeper, it was realised that the true period is 545 secs, and that the pulsation is double-peaked. The data, folded on the 545-sec period, are by Alasdair Allan.[20]

two peaks. As discussed previously, occultation effects at opposite poles cancel out, while absorption effects in the accretion curtain model produce a single-peaked pulse. Note that EX Hya has a single-peaked pulse even though the partial X-ray eclipse proves that we see both poles.

Instead, we have to consider how the radiation is beamed by the accretion region. The accretion curtain model suggests that curtains appear brightest when they display maximum surface area, so that most radiation emerges perpendicularly to the sweep of the curtain (that is, perpendicular to both the infall direction and the direction in which the accretion arc is extended). Alternatively, if the opposite of the accretion-column idea were the case, so that the accretion regions were brightest when we looked directly at them, then the two poles, alternately flitting across the face of the white dwarf, would combine to produce a double-peaked pulsation. Currently, we do not have enough observational clues to determine why some systems show double-peaked pulses while others are single peaked.

9.4.3 Optical pulsations at the beat period

Whereas the X-ray pulsations in disc-fed intermediate polars are predominantly at the spin period, the optical lightcurves often show a strong beat-cycle pulsation. Indeed, in some systems the optical lightcurve is dominated by a beat-cycle pulsation, with no spin pulse being detectable (see Fig. 9.17).

Beat-cycle pulsations arise through the interaction of the spin and orbital cycles, the particular mechanism in this case being the irradiation of the binary by the pulsed X-rays. When the accretion flow absorbs X-rays it heats up, increasing its luminosity in the optical. Absorption by a symmetric structure, such as an accretion disc, or by structure locked to the white dwarf, such as the accretion curtains, will simply result in an optical spin pulse. However, irradiation of structure locked to the binary orbit, such as the bright spot where the stream hits the disc, will instead produce a beat-cycle pulsation. This arises since the orbital motion of the bright spot means that a rotating beam of X-rays must travel slightly further than

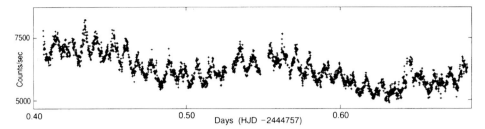

Fig. 9.17: The optical lightcurve of V1223 Sgr is dominated by a pulsation at the 794-sec beat period, not the 745-sec spin period. Also obvious is a smooth modulation on the 3.37-hr orbital cycle. (Data by Mark Cropper and Brian Warner.[21])

one revolution before re-illuminating the bright spot, and this again leads to the mathematical relation of Box 6.1.[††] The simple interaction of the two cycles to produce a third can be further complicated to produce a whole range of orbital sidebands of the fundamental pulsation, many of which are weakly detected in intermediate polars. This is discussed further in Box 9.2.

9.4.4 DQ Her

Before leaving this section, DQ Her itself deserves a mention. Despite being the first intermediate polar identified, it violates the main membership criterion by not showing an X-ray pulsation, nor indeed much X-ray flux at all. On closer examination, however, it turns out to be the exception that probes the rule.[‡‡] DQ Her is seen almost exactly edge on, at an inclination of 87°, so that the thick accretion disc completely hides the white dwarf, blocking our view of the X-ray emission. But we know that there is a pulsed X-ray flux, since the disc sees it. We deduce this from a 71-sec optical pulsation in the lightcurve: by watching the pulsations gradually die away as the accretion disc enters eclipse (Fig. 9.18), we deduce that they come from the irradiated inner surface, which periodically brightens as the X-ray beam sweeps by.[22]

9.5 PROPELLERS

There is one further type of behaviour in which a magnetic cataclysmic variable can indulge: acting as a propeller, and expelling the material that attempts to accrete. This occurs if the spin period is much faster than the equilibrium period for the white dwarf's field strength (or, in an equivalent statement, if the magnetospheric radius is larger than the corotation radius).

In this situation the rapidly spinning field lines, whipping across the surface of the diamagnetic blobs, increase the energy of the blobs, expelling them to larger distances. This can be sufficient to drive the blobs out of the binary altogether.

---

††Note that X-ray beat pulses cannot be produced in this way, since most of the X-rays are degraded to cooler radiation.
‡‡This wording makes the original meaning of the remark clearer.

Box 9.2: Multiple orbital sidebands in intermediate polars

The fundamental spin and orbital cycles (with frequencies ω and Ω respectively) change their relative orientation at the beat frequency $\omega - \Omega$. However, as first pointed out by Brian Warner, more complex interactions between the two cycles give rise to additional 'orbital sidebands' of the spin pulse.[23]

Consider a spin pulse whose amplitude changes over the orbital cycle (if disc structure periodically obscures the white dwarf, for example). This could be represented by the function $\sin \omega t \cos \Omega t$ (where t is the time and ω and Ω are angular frequencies, that is $2\pi f$ where f is in hertz). A standard trigonometric relation then gives

$$\sin \omega t \cos \Omega t = \frac{1}{2}[\sin(\omega - \Omega)t + \sin(\omega + \Omega)t].$$

Thus the effect of the amplitude modulation is to put power equally into the two sidebands $\omega - \Omega$ and $\omega + \Omega$. Furthermore, if the amplitude modulation is more complex than a $\cos \Omega t$ function, it could be represented by the series $\cos \Omega t + \cos 2\Omega t + \cos 3\Omega t \ldots$ which would produce sidebands of $\omega \pm \Omega$, $\omega \pm 2\Omega$, $\omega \pm 3\Omega$, etc (to keep things simple I am ignoring the arbitrary amplitudes and phasings of the different components).

The same process can affect the beat pulse. Suppose a $\omega - \Omega$ pulse is created by X-ray irradiation of the bright spot. The amplitude of the $\omega - \Omega$ pulse would be different at orbital phases when the irradiated side of the bright spot faced towards us compared to phases when it faced away from us. Thus the $\omega - \Omega$ pulse would be amplitude-modulated at Ω, putting power into the sidebands of $\omega - \Omega$, namely $\omega - 2\Omega$ and ω. By involving harmonics of both periods, more extensive sets of sidebands are, in principle, possible.

The Fourier transform below, of the optical lightcurve of FO Aqr, shows the set of sidebands predicted by Warner's theory (figure by Joe Patterson[11]).

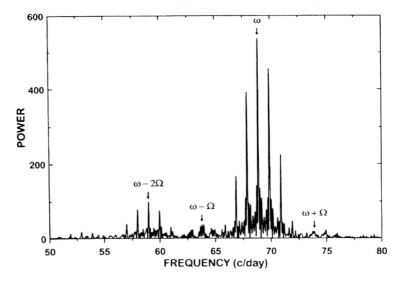

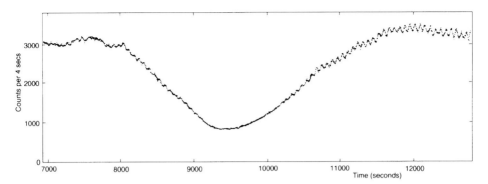

Fig. 9.18: The 71-sec optical oscillation in DQ Her arises from reprocessing of X-rays by the accretion disc. The pulsation thus dies away gradually as the disc is eclipsed. (Lightcurve by Ed Nather.[24])

9.5.1 AE Aqr

We have at least one candidate for a propeller system in AE Aqr — a magnetic cataclysmic variable with a host of peculiarities. Its unusually long orbital period, at 9.9 hrs, and its unusually short spin period, at 33 secs, would imply a small magnetosphere surrounded by a large disc, if the system were in equilibrium with its accretion.

However, the mass transferred by the secondary star (judged using the optical

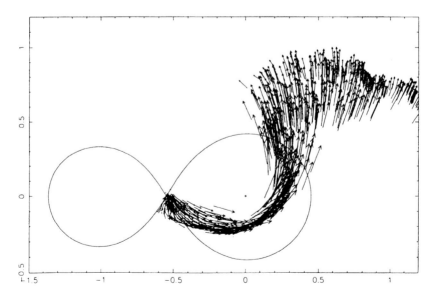

Fig. 9.19: A model of a 'propeller' system in which the rapidly spinning magnetic field expels the accretion stream from the system. (Figure by Graham Wynn.[25])

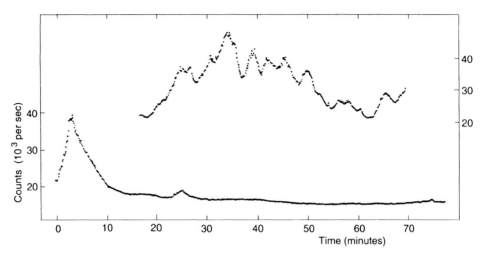

Fig. 9.20: AE Aqr has a distinctive optical lightcurve, with episodes of strong 'flaring' and episodes of near quiescence. (Lightcurves by Joe Patterson.[26])

light from the accretion flow) appears to be 1000 times greater than the mass actually accreting onto the white dwarf (as judged from its feeble X-ray flux). This suggests that the rapidly spinning field is ejecting the majority of the flow from the binary (see Fig. 9.19). The exertion of expelling the flow causes the white dwarf to lose energy, decreasing the rate at which it spins. Indeed, the white dwarf is losing rotational energy at 100 times the rate at which the accretion flow radiates.

Other peculiarities include the unusual 'flaring' lightcurve of AE Aqr (Fig. 9.20), which could result from collisions between diamagnetic blobs, and the possibility that AE Aqr emits ultra-high-energy gamma rays, in the TeV range.

To complete the picture we need to explain why AE Aqr has a spin period much faster than its equilibrium value. The hypothesis is that it previously underwent an episode of much higher accretion rate ($\dot{M}$), when a dense disc pushed the magnetosphere inwards to smaller radii, spinning it up until it corotated with the inner disc. Then, after a drop in $\dot{M}$, the disc dissipated. At the lower $\dot{M}$ seen today, the field is spinning too rapidly for the larger magnetosphere, and so acts as a propeller. Over time, given the energy drain of the propeller, the spin rate will decrease until disc-fed accretion can resume.[25]

This hypothesis implies that the mode of accretion seen today can be determined as much by the history of the system as by the basic parameters such as orbital period and field strength.

9.5.2 WZ Sge

A propeller might also be operating in WZ Sge, judged by the detection of pulsed X-rays at a period of 28 secs. If interpreted as a white-dwarf spin period, this is faster than in any other magnetic cataclysmic variable.

The previous discussion of WZ Sge's outbursts left a puzzle that a propeller might resolve. The 33-yr intervals between superoutbursts are not filled with normal outbursts. In standard disc-instability models this can only be produced by invoking an exceptionally low viscosity in quiescence, so that the disc material does not diffuse into the inner disc and trigger an outburst.

Instead, a magnetic propeller could expel material from the inner disc, preventing the build-up that would trigger an outburst.[27] Material then accumulates in the outer disc, where the mass required to trigger an outburst is much larger. Thus the inter-outburst interval is lengthened, and, when outbursts do occur, they involve sufficient material to become superoutbursts. The inward surge of material in a superoutburst overwhelms the magnetic field, collapsing the magnetosphere onto the white-dwarf surface, and allowing accretion to occur. Hence WZ Sge shows superoutbursts, but no normal outbursts.

9.6 EVOLUTION OF MAGNETIC CATACLYSMIC VARIABLES

The evolution of most intermediate polars is probably little different from that of non-magnetic systems, since evolution is driven by the secondary star, far from the influence of the white dwarf's magnetic field. However, as it evolves to shorter orbital periods the magnetosphere fills more and more of the binary. Once the orbital separation has shrunk to be comparable to the magnetospheric radius, the interlocking fields of the two stars could overcome the accretion torques, synchronising the system and turning it into an AM Her star.

In fact, the range of field strengths of the polars (~ 10–100 MG) overlaps somewhat with those of intermediate polars (~ 1–10 MG). Thus the highest-field intermediate polars are PQ Gem (≈ 15 MG with a 5.2-hr orbit) and V2400 Oph (≈ 15 MG with a 3.4-hr orbit), while the lowest-field polars are EF Eri (8 MG with a 1.4-hr orbit) and AM Her itself (13 MG with a 3.1-hr orbit). Almost certainly these high-field intermediate polars will become polars once they evolve below the period gap. However, most intermediate polars do not show polarisation, and probably have fields nearer 1 MG. These will not synchronise to become polars.

The evolution of AM Hers is less clear. Since evolution is driven by the magnetically threaded wind of the secondary, the interlocking of field lines from the primary with those of the secondary is expected to have a major effect. However, we as yet have no settled account of evolution under these circumstances.[28]

Box 9.3: The accretion column in intermediate polars

The accretion columns in intermediate polars are similar to those in AM Her stars (Section 8.6), but there are also differences. One major difference is that accretion in intermediate polars is less 'blobby' than in AM Hers, since the process of stripping material from a disc leads to diffuse material, which forms a hard-X-ray emitting shock. Thus the copious soft-X-ray production of AM Hers is not seen. However, soft blackbody emission with temperatures of ~ 50 eV has been detected in 3–4 intermediate polars, at a level appropriate for 10^{-4}–10^{-5} of the white-dwarf surface; this is probably the heated region surrounding the main accretion column. This emission is not seen in other systems, but may be being absorbed by the accretion flow or by interstellar material.

Another difference from AM Hers is that the accretion footprints are expected to be longer arcs, because an accretion disc would feed field lines over a greater range of azimuth (though the extent of this is unclear). Counterbalancing this is the possibility that disc-fed accretion produces a footprint with a smaller width, since feeding of field lines might occur over a small range of radii; in AM Hers the stream punches a hole in the magnetosphere and stripping of the stream could occur over a greater radial range.

The lesser importance of cyclotron emission in an intermediate polar simplifies the calculation of the temperatures and densities in the column. As in an AM Her, the infalling material encounters an accretion shock some distance above the white-dwarf surface. Here the infall velocity drops by a factor 4 and most of the energy is converted to heat. The variation in temperature and density as the material cools by bremsstrahlung emission and settles onto the white dwarf can be calculated from hydrodynamical equations, and is shown below (assuming 10^{13} kg s^{-1} falling onto 10^{-3} of the surface area of a 0.6-M$_\odot$ white dwarf). Given the temperature and density at each point one can calculate the expected X-ray emission from the hot plasma, and integrate to find the total X-ray emission from the column. Of course, this then has to be modified by the absorption encountered as it emerges through the accretion flow.[29]

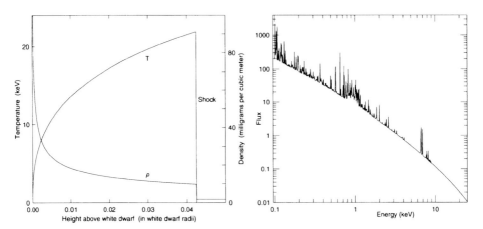

Left: The temperature and density through the accretion shock. *Right:* The X-ray spectrum resulting from such a column. (Computations by Andrew Beardmore.)

Chapter 10

Flickering and oscillations

Much of this book has focused on periodic variations in the lightcurves of cataclysmic variables, which recur like clockwork with the binary orbital period or the white-dwarf spin period. However the most striking feature of a typical lightcurve is how much random variability there is, which does not recur with any cycle. This is called *flickering*, and is illustrated in Figs 10.1 and 10.2.

Flickering occurs on many timescales, from rapid fluctuations lasting a few seconds to larger flares and dips lasting for hours. It reminds us that transfer of material from the red star to the white dwarf is a turbulent, not a smooth process. The only pattern that can be discerned in flickering is that the longer-lasting fluctuations have larger amplitudes. In an analogy with the frequency distribution of coloured light, this is termed 'red noise' (whereas 'white noise' has equal amplitudes at all frequencies). Red noise is expected from a turbulent flow since the swirling motion in a larger eddy will take longer, and the larger eddy will emit more light.

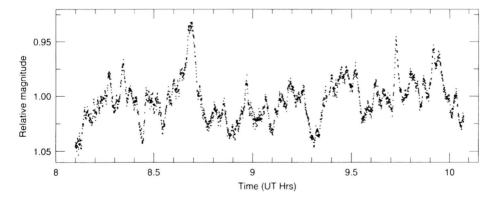

Fig. 10.1: Flickering in the Z Cam star RX And. The variability has no pattern, does not repeat, and has no preferred period, although the longer-lasting fluctuations have larger amplitudes. The data are 3-sec integrations taken on 1976 August 20 at McDonald Observatory. (Lightcurve by Ed Nather.)

152 Flickering and oscillations

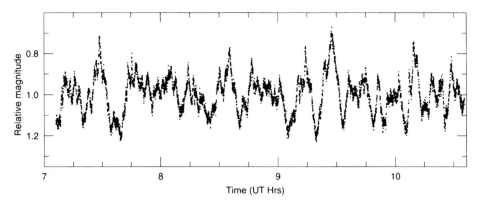

Fig. 10.2: Flickering in the novalike KR Aur. The data are 2-sec integrations taken on 1977 November 14 at McDonald Observatory. (Lightcurve by Ed Nather.)

10.1 THE LOCATION OF FLICKERING

An influential lightcurve of U Gem taken in 1971 (Fig. 10.3) showed that the flickering disappeared during the eclipse. Since the grazing eclipse in U Gem occults the bright spot but not the central disc (see Section 2.4.3), this implied that the flickering arose from the bright spot, caused by the turbulent impact of the accretion stream. However, later lightcurves have shown the reverse: for instance in HT Cas, where the eclipses of both bright spot and white dwarf are seen in the lightcurve, flickering reappears when the white dwarf has emerged from eclipse, and the bright spot is still hidden, thus demonstrating that the flickering arises in the inner disc close to the white dwarf.

More rigorously, one can construct a 'flickering lightcurve', observing many cy-

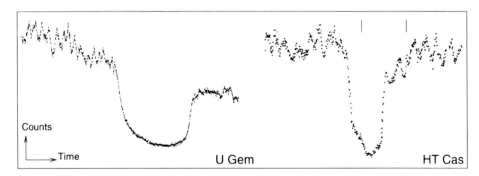

Fig. 10.3: A tale of two eclipses. U Gem shows a grazing eclipse of the bright spot only, thus the disappearance of flickering in eclipse indicates that it originates in the bright spot. HT Cas shows the classic two-stepped eclipse of a white dwarf and bright spot (cf. Fig 2.11). Since flickering resumes before the egress of the bright spot (whose ingress and egress are marked by two ticks) it must originate in the inner disc. [Data by Brian Warner (U Gem) and Joe Patterson (HT Cas)[1].]

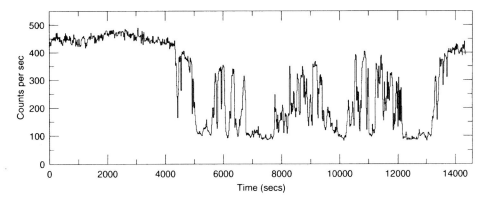

Fig. 10.4: An X-ray lightcurve of the low-mass X-ray binary X1624–49. The prominent 'dip', which lasts for more than 2 hours out of the 22-hr binary period, is caused when the bright spot obscures the view of the neutron star, absorbing the X-rays (see Fig. 7.7). The extreme, rapid variability within the dip implies that the stream–disc impact is highly turbulent, and so will be a source of optical flickering. The data are 16-sec samples of the 2–20-keV lightcurve observed with *RXTE*. (Lightcurve courtesy of Alan Smale.)

cles of a star and calculating the average deviation from the mean lightcurve at each orbital phase. Where this has been done it shows a reduction in flickering centred on the eclipse of the disc.[2] Thus flickering appears to arise predominantly in the turbulent inner disc, but also, in at least some systems, from the bright spot.

10.2 QUASI-PERIODIC OSCILLATIONS

By definition, flickering is not periodic. However, sometimes the variability does occur on a preferred timescale, to produce *quasi-periodic oscillations*. These are not the strictly periodic oscillations seen in intermediate-polar lightcurves, but are oscillations that last for only a few cycles, then die away to be replaced by further oscillations with a different phase or with a slightly different period. Typically such oscillations (called QPOs for brevity) have amplitudes of a few percent and periods between 30 and 300 seconds (though longer-period QPOs are harder to detect and may have been overlooked).

The QPOs appear to originate in the accretion disc, and one can speculate about blobs of material orbiting at the local Keplerian velocity, circling a few times before being subsumed into the disc. Alternatively, a perturbation of the disc, either radially or out of the plane, could lead to an oscillation at the local Keplerian velocity; adjacent annuli would act independently and with slightly different frequencies, thus producing a quasi-periodic phenomenon. Although such theories are plausible, there is as yet insufficient evidence to prove or disprove them.

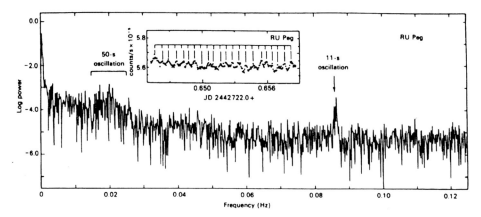

Fig. 10.5: This Fourier transform of RU Peg's lightcurve reveals an 11-sec DNO and a less-coherent QPO at 50 secs. The inset shows the lightcurve, where the maxima of the 50-sec oscillation are marked. (Figures by Robinson and Nather,[3] as adapted by Warner.)

10.3 DWARF-NOVA OSCILLATIONS

Another type of non-periodic oscillation has become known as a *dwarf-nova oscillation*, or DNO for short. DNOs are only seen during the outburst of dwarf novae, have periods of 7–30 secs (shorter than the QPOs), and are far more coherent than QPOs, being able to maintain a stable phasing for an hour or more. Occasionally, DNOs are so strong that they can be seen directly in a lightcurve (such as that in Fig. 10.6), although more often they are revealed by Fourier analysis.[4]

The DNO typically appears on the rise to outburst, then evolves to a shorter period, having a minimum value near the outburst peak, and then lengthens its period before disappearing on the outburst decline (Fig. 10.7). This behaviour suggests that higher accretion rates result in shorter periods.

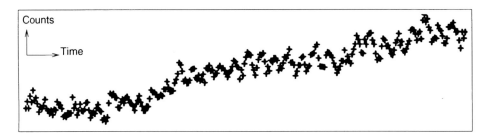

Fig. 10.6: An optical lightcurve of TY PsA in which DNOs with a period of 27 secs can be seen directly. The data shown cover 900 secs. (Adapted from work by Warner, O'Donoghue and Wargau.[5])

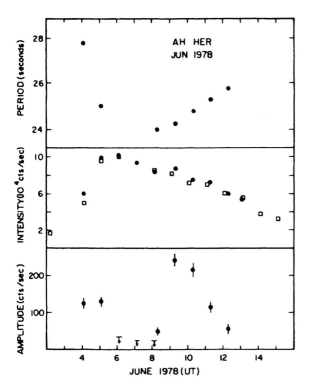

Fig. 10.7: The variation of DNOs over a 10-day outburst of AH Her. The panels shows the DNO period (*top*), the star's brightness (*middle*), and the DNO amplitude (*bottom*). (Figure by Joe Patterson.[4])

10.3.1 Weakly magnetic white dwarfs?

The most promising explanation of DNOs is that they are similar to the pulsations seen in intermediate polars, with the difference that the white dwarfs have weaker magnetic fields, say $\sim 10^5$ G, able to channel the accretion flow only very near the white dwarf.[6] The DNO periods, being generally shorter than those in intermediate polars, then fit the trend of faster white-dwarf rotation with weaker fields. Indeed, the periods are close to the Keplerian period of disc orbits just above the white-dwarf surface.

Note, however, that the DNO periods show large changes, of up to $\sim 25\%$, over an outburst. Given the huge inertia of a white dwarf it is clearly unable to change period this rapidly. Thus the DNO must involve only a small fraction of the white dwarf, say $10^{-10}\,M_\odot$. This is probably the surface layer of the equatorial belt at the boundary layer, spun up by the increased accretion torques in outburst. This can occur with a weak-field white dwarf, whereas a stronger field would anchor the outer layers to the core, preventing any slippage.

The main difficulty with this model is explaining why DNOs are seen only in outburst. The standard theory of magnetic accretion (Box 8.1) suggests that the lower accretion rates of quiescence would lead to a larger magnetosphere and thus better channeling of accretion. But DNOs have never been seen in quiescence. In particular, SS Cyg in outburst shows soft-X-ray emission containing DNOs with

156 Flickering and oscillations   Ch 10

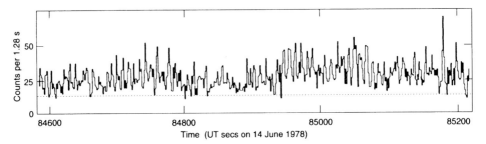

Fig. 10.8: The soft-X-ray lightcurve of SS Cyg showing 9-sec DNOs. The data are from the *HEAO*-1 satellite in the 0.1–3-keV range, binned to 1.28 secs. The dotted line is the estimated background level. (By France Córdova and colleagues.[7])

a period of 7–12 secs. In quiescence it shows hard X-rays which are not pulsed. Perhaps the best explanation is that the field is not there in quiescence, but is generated in outburst by a dynamo created by the shear between the equatorial belt of the white dwarf and the layers beneath.

10.3.2 Beat frequencies from the magnetospheric boundary?

A magnetic accretor also provides potential for quasi-periods arising at the boundary between the magnetosphere and the disc. Although in equilibrium the Keplerian velocity at the inner edge of the disc will match the white dwarf's rotation rate, in practice systems can be out of equilibrium. The interaction between a blob of material at the inner edge of the disc and the magnetic field would then vary with the beat frequency between the Keplerian and spin frequencies — the frequency at which a particular field line encounters the blob as they chase each other round.* A variation in the viscous interaction or the accretion rate with this frequency would produce a modulation in the optical or X-ray lightcurves. Since the magnetospheric radius, and thus the Keplerian frequency at that radius, would fluctuate according to variations in the mass-transfer rate, the beat period would be variable.

It is thus possible that DNOs occur at the beat period, not the spin period. Indeed, this can explain the period changes in outburst: the increased mass-transfer of outburst will drive the magnetosphere inwards and out of equilibrium, causing the beat period to become shorter, as seen; when the accretion rate eases at the end of outburst the beat period lengthens again.

It is presently unclear whether this mechanism is responsible for DNOs, although it probably occurs in some intermediate polars, where optical quasi-periods close to the spin period are sometimes seen, and which almost certainly indicate a magnetosphere spinning slightly out of equilibrium.[8]

---

*The usual relation applies (cf. Box 6.1): $1/P_{\rm beat} = 1/P_{\rm spin} - 1/P_{\rm Keplerian}$.

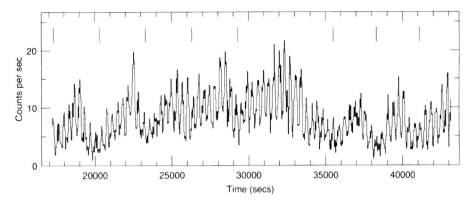

Fig. 10.9: In outburst the X-ray lightcurve of GK Per showed a prominent 351-sec pulsation, and semi-regular 'dips' recurring with a timescale of 3000–5000 secs. Data from the 1983 outburst in the 2–8-keV range observed with *EXOSAT*.

10.4 GK PER

GK Per is an intermediate polar with a 351-sec spin period that also undergoes dwarf-nova outbursts. It has also shown weak optical quasi-periodicities near 380 secs. During an outburst in 1983 an X-ray lightcurve showed prominent 'dips' on a semi-regular timescale of 3000–6000 secs (see Fig. 10.9).[9]

One plausible explanation for this is that 380 secs was the Keplerian period of the inner disc, leading to a ~ 5000-sec beat period. A modulation of the accretion rate at this beat period would then produce the dips. However, this implies that the accretion rate varied by factors of 3 on hourly timescales, which is unlikely. Furthermore, the X-ray dips were deeper at lower energies, characteristic of absorption, which suggests that the dips resulted from obscuration of the white dwarf by structure in the disc (see Section 7.3 for similar behaviour). Thus a variant of this idea is that the interaction of the field with blobs in the inner disc allowed extra material to follow field lines out of the disc plane, causing extra absorption which was modulated at the beat period. However, this is not in accord with what was observed: the absorbing blobs would still be orbiting with a period close to 380 secs, and so would be on the far side of the white dwarf for roughly half the spin cycle, where they could not absorb X-rays. But the dips clearly reduce the X-ray flux at all phases of the 351-sec spin cycle.

This has led to speculation that the dips arise from a different mechanism, where the absorbing material is further out, and so is actually orbiting at the ~ 5000-sec dipping timescale. It is notable that if the accretion stream overflows the bright spot and re-impacted the inner disc (as discussed in Section 7.3), it would re-impact in GK Per at a radius where the Keplerian period was close to the observed dipping timescale. Thus it is possible that the stream re-impact produces structure orbiting at this radius that periodically obscures the white dwarf, producing dips.[10] As usual with models of quasi-periodicities, we as yet have neither proof nor disproof.

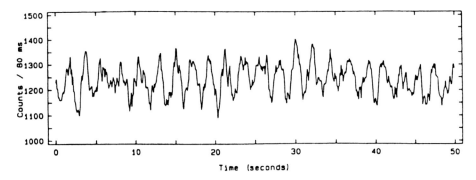

Fig. 10.10: A lightcurve of the AM Her star V834 Cen in which QPOs are visible directly. (Figure by Stefan Larsson.[11])

10.5 QPOS IN AM HER STARS

The magnetic AM Her stars produce their own brand of QPOs, which arise due to an instability in the accretion shock formed as the infalling material approaches the white dwarf (such shocks were described in Section 8.6). The instability can be understood as follows. Suppose the shock-front has a small upward movement; this increases its velocity relative to the incoming material, which increases the energy dissipated in the shock and thus heats it up. The hotter material cannot cool sufficiently over the length of the column, so the excess heat produces a pressure which pushes the shock-front even higher. Thus upward movements are magnified and, by the reverse argument, downward movements are similarly magnified. This explains why a shock is unstable, and has a natural tendency to oscillate up and down with periods of a few seconds. However, something still needs to trigger the initial perturbation, and this is done by the random arrival of 'blobs' of accretion.[12]

If a single oscillation caused by a single blob dominates, it can be strong enough to be seen directly in optical lightcurves (Fig. 10.10). However, more usually there are several blobs acting independently, exciting simultaneous oscillations with different phases and slightly different periods. Fourier analysis is needed to detect such QPOs, which appear as broad humps of power, rather than individual frequencies (Fig. 10.11).

10.6 PULSATING WHITE DWARFS

The flickering and oscillations discussed so far in this chapter all originate in the accretion flow between the two stars. To complete the picture, we should also consider pulsations of the white dwarf itself. The subject of pulsations in stars is vast, and merits a book-length treatment in its own right. Since it is peripheral to cataclysmic variables, though, I will deal with it only briefly. Essentially, waves of pressure can run through stars, causing them to vibrate at their natural frequency, in the same way that bells ring with a characteristic note. The frequency depends

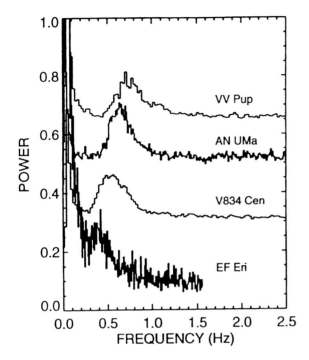

Fig. 10.11: Fourier transforms showing QPOs in four AM Her stars at frequencies between 0.3 and 1.0 Hz. (Figure by Stefan Larsson.[13])

on the density of the star,[†] so that large, diffuse stars such as Cepheids and Mira variables pulsate on timescales of tens or hundreds of days, respectively, whereas small, dense white dwarfs pulsate with typical periods of a few hundred seconds.

As with bells, the pulsation would die away unless there was a driving force. This is provided by energy leaking out from the hot core of the white dwarf. Normally, the energy can leak out steadily, but if the material is partially ionised a 'valve' mechanism forces it to emerge in spurts. This arises because the partially ionised material, when compressed, becomes more opaque to radiation, entrapping more of the energy. The compressed material thus gains energy, causing it to expand until the energy can flow out, whereupon it contracts to undergo another cycle.

In a white dwarf with a surface temperature of $\sim 12\,000$ K, critical layers of hydrogen are partially ionised, driving pulsations. Such stars are known as 'ZZ Ceti' stars. It had long been though that accretion in cataclysmic variables would keep their white dwarfs too hot to be ZZ Ceti stars, but in 1997 Brian Warner and Liza van Zyl discovered that the white dwarf in the 18th-magnitude dwarf nova GW Lib was pulsating.[14]

The hallmark of such a system is a set of pulsations too complicated to be explained by any magnetic-accretion hypothesis. Whereas a one-dimensional guitar string can vibrate at its fundamental frequency and at a series of higher 'overtones', depending on the number of wavelengths along its length, the two-dimensional surface of a white dwarf can vibrate in many more modes, depending on the number

---

[†]Dimensional analysis shows that the period varies $\propto (G\rho)^{-1/2}$.

160   Flickering and oscillations   Ch 10

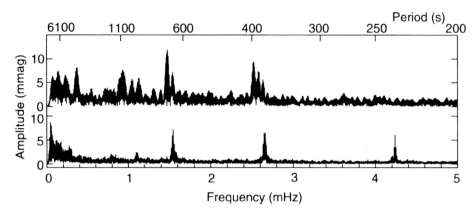

Fig. 10.12: Fourier transforms of the optical lightcurve GW Lib, obtained in April 1997 (*upper*) and May/June 1998 (*lower*). The data reveal pulsations near 236, 376 and 650 secs, each made up of several components. Note also the changes between the two epochs, such as the absence of the 236-sec pulsation in the earlier data. The array of pulsations and their variability are characteristic of pulsating white dwarfs. (Data from the *Whole Earth Telescope*, courtesy of Liza van Zyl.[15])

of wavelengths running from the north pole to the south pole, and the number running along the equator. A Fourier transform of a lightcurve of GW Lib (Fig. 10.12) shows three main pulsations at 236, 376 and 650 secs. However, closer inspection reveals that each of these is a complex of several pulsations, such that at least 15 overtones are needed to explain the lightcurve!

GW Lib is a WZ Sge-type dwarf nova, with only one known outburst, in 1983. The low accretion rate of such systems may account for the white dwarf being cool enough to pulsate, so it is worth looking for pulsations in similar dwarf novae. Further, GW Lib presents an opportunity to study how cooling and heating over the outburst cycle affects pulsations. If the white dwarf is still cooling from the 1983 outburst, the changes in the pulsations may account for some of the complexity of its lightcurves.

Chapter 11

The nova eruption

The most dramatic event in the life of a cataclysmic variable is the nova explosion. Since antiquity Chinese astronomers had recorded 'guest stars', appearing for a few months before fading, while in the West they were called *novae stellar* or 'new stars'. In the last century a score of novae were bright enough to be easily visible to the naked eye, and one, V603 Aql, which erupted in 1918, was for a time the second-brightest star, outshone only by Sirius.

Whereas dwarf-nova outbursts have typical amplitudes of 3–5 mags and repeat every few months, nova eruptions have typical amplitudes of 8–15 mags and are not seen to repeat.* Few novae have been caught on the rise to eruption, but those that have typically brighten over 1–3 days. The speed of decline is commonly measured as the time taken to decline by three magnitudes from the peak. V1500 Cyg, one of the fastest of recent novae, took 3.6 days; HR Del, one of the slowest, took 230 days (both curves are shown in Fig. 11.1). The faster novae have higher amplitudes (those of V1500 Cyg and HR Del were 14.5 and 8.5 mags respectively).

11.1 THERMONUCLEAR RUNAWAYS

The cause of a nova eruption is a nuclear chain-reaction on the surface of the white dwarf. The high temperature and pressure at the surface of a white dwarf cause the layer of accreted material to explode like a hydrogen bomb — but this hydrogen bomb is 30 times the mass of the Earth. The source of the energy is the strong binding of protons and neutrons into nuclei. Hydrogen nuclei (lone protons) have no binding energy, but burning 1 kg of hydrogen into helium releases 6×10^{14} J — enough energy to power the United Kingdom for a day.

The white dwarf itself is the burnt-out core of a star, in which most of the hydrogen has been burnt into helium and heavier elements such as carbon, oxygen, neon and magnesium. Accretion from the secondary, however, is essentially a supply of fresh hydrogen, which can restart nuclear reactions. As a layer of accreted hydrogen builds up, its weight, in the extreme gravity of the white dwarf, crushes

---

*A class of recurrent nova is described in Section 11.4; their amplitudes and recurrence times can overlap with those of the WZ Sge-type dwarf novae.

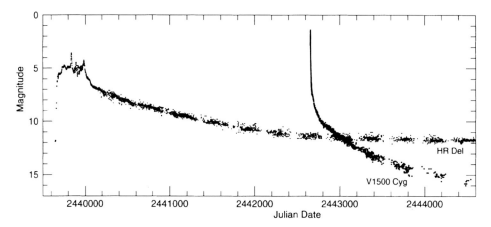

Fig. 11.1: The lightcurves of HR Del (Nova Del 1967) and V1500 Cyg (Nova Cyg 1975), representing the extremes of fast and slow novae. The data are 1-day averages compiled by the AAVSO.

the material at the base of the layer. At such densities the material is degenerate, with the electrons being pushed to occupy the same space, which is forbidden by Pauli's exclusion principle. This results in a degeneracy pressure, keeping the electrons apart, but only by forcing the electrons into states of higher energy. As more accretion increases the pressure further, the temperature and density increase sufficiently to cause nuclear reactions.

The first reaction is a collision of two protons in which one proton changes into a neutron by the emission of a positron. This forms a 'heavy hydrogen' nucleus called a deuteron. The deuteron can then absorb another proton to become helium. In normal matter the energy released by such reactions would increase the pressure, causing the material to expand and cool, and thus moderating the reaction. In degenerate matter, however, the pressure is determined solely by the density; the temperature increase does not increase the pressure and doesn't cause expansion. It does, though, increase the rate of nuclear reactions, leading to a runaway effect as reactions increase the temperature which increases the reaction rate, and so on.

When the temperature has risen to 2×10^7 K, hydrogen can burn more efficiently by using nuclei of carbon, nitrogen and oxygen as catalysts (see Box 11.1). These elements absorb protons, resulting in nuclei which are radioactive with half-lives of ~ 100 secs. Such reactions are exquisitely dependent on temperature (with a rate $\propto T^{18}$) so the runaway speeds up as the temperature climbs past 10^8 K. The huge energy input due to nuclear reactions at the base of the accreted layer drives the layer into convective motion, sending radioactive nuclei to the surface and sucking down fresh material for burning.[1]

The runaway continues until the rise in temperature causes the gas pressure to exceed the degeneracy pressure. This relieves the degeneracy, and the layer now expands. The radioactive nuclei deposit their energy into the expanding shell, driving further expansion. At ~ 1000 secs after ignition, the shell can be radiating

Box 11.1: The reactions of a thermonuclear runaway

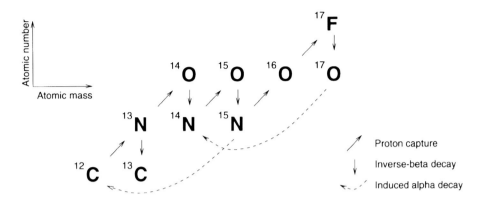

In the nuclear powerhouse of a nova eruption, nuclei of carbon, nitrogen and oxygen catalyse the burning hydrogen into helium, in a 'CNO cycle'. The cycle proceeds by proton captures, such as

$$p + {}^{12}C \rightarrow {}^{13}N + \gamma,$$

coupled with inverse-beta decays, such as

$$^{13}N \rightarrow {}^{13}C + e^+ + \nu_e,$$

and is completed by proton captures that induce alpha decay, for example

$$p + {}^{15}N \rightarrow {}^{12}C + {}^4He.$$

The overall result is the absorption of four protons (hydrogen nuclei) and the emission of one alpha particle (helium nucleus). Meanwhile, the abundant radioactive isotopes ^{13}N, ^{14}O, ^{15}O and ^{17}F, which inverse-beta decay on a ~ 100-sec timescale, release the energy to drive the expansion of the nova shell.[1]

---

at 100 000 solar luminosities, mostly due to radioactivity, and expanding outwards at ~ 3000 km s^{-1}.

11.2 THE EXPANDING NOVA SHELL

The hot shell expands to engulf the whole binary. The binary motion now acts as a propeller, helping to eject the shell (effectively a temporary return to the 'common envelope' phase from which the cataclysmic variable formed). At this stage the spectrum of the nova shows blue-shifted absorption lines (Fig. 11.2), the P Cygni profiles characteristic of an outflow (see Section 7.2.1). From these lines we deduce that the shell flows outwards at speeds between ~ 500 km s^{-1} (typical of a nova

164 The nova eruption

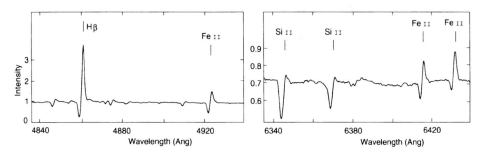

Fig. 11.2: Regions of the spectra of Nova Cas 1995 (V723 Cas) taken on 1995 October 13, two months after the nova ignition. The prominent blue-shifted absorption features in the lines are the classic 'P Cygni' signature of an outflowing wind. (Adapted from work by Iijima, Rosino and della Valle.[2])

with a slow decline) and ~ 2000 km s^{-1} (typical of a nova with a fast decline). As the shell expands and cools, the nova fades.

After ~ 1000 days (depending on the distance to the nova) the shell will have become big enough to be seen as nebulosity surrounding the binary (Fig. 11.3). This provides a very useful method of determining the nova's distance. By measuring the expansion speed from Doppler shifts in the spectrum, and using the time since the eruption, one can calculate the absolute size of the shell. By comparing this with the apparent size one deduces the distance.

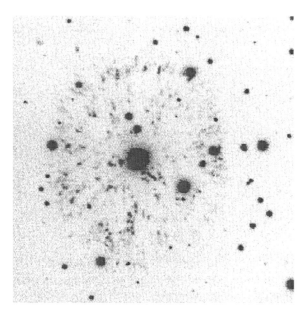

Fig. 11.3: A CCD image of nebulosity surrounding GK Per (the bright central star), a relic of the shell of material ejected in its nova eruption of 1901.

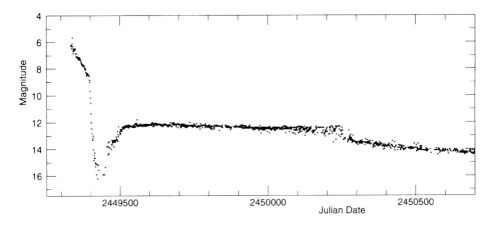

Fig. 11.4: Nova Cas 1993 (V705 Cas) showed a deep 'dust minimum' caused by the nova becoming shrouded by dust condensing out of the nova shell. The data are 1-day averages compiled by the AAVSO.

One further feature of the expanding shell should be mentioned, since it can have a dramatic effect on the nova lightcurve. This is the formation of 'dust' in the outflow. Once the temperature in the shell drops to ~ 1000–2000 K, atoms of carbon and silicon condense into tiny grains, typically 0.01–1 μm across. The 'dust' of such particles blocks the radiation from the nova, and its optical luminosity plummets (Fig. 11.4). The onset of dust formation is rapid, since once grains start to form they block the radiation, allowing more grains to condense. The trapped energy is absorbed by the grains, and reradiated at the infrared wavelengths appropriate for the large, cool shell. As the shell continues to expand, however, the dust dissipates and the optical radiation can again escape, leading to an increase in the observed brightness.

It is possible that dust formation might be responsible for the strong oscillations seen during the decline of some novae (most recently in V1494 Aql, see Fig. 11.5). This could occur if dust traps enough energy to destroy the grains, which then allows the energy to escape, leading to renewed dust formation. However, such oscillations are poorly studied, so their origin is uncertain.

Over time, the nova shell dissipates into the interstellar medium, and after a few hundred years it is no longer seen.

11.3 THE RECURRENCE OF NOVA ERUPTIONS

The main factor in determining the speed of a nova is the mass of the white dwarf, the more massive systems generating the faster eruptions. The white-dwarf mass also determines the amount of accreted material that accumulates before triggering a nova eruption. The smaller mass and larger radius of a lighter white dwarf combine to give a much lower surface gravity, which needs a larger accreted layer to produce the pressure required to trigger an explosion. Calculations suggest that a 0.6-$M_\odot$

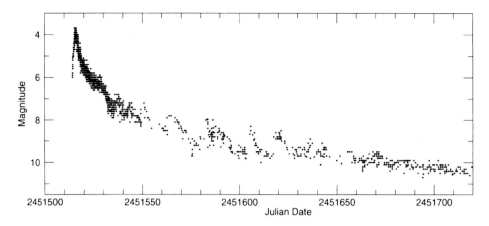

Fig. 11.5: During its decline, Nova Aql 1992 No. 2 (V1494 Aql) showed oscillations with an amplitude of over 1 mag. At least four brightenings, and possibly more, were seen recurring at ∼ 20-day intervals. (Data from the AAVSO and VSNET.)

white dwarf needs to accrete $\approx 5\times 10^{-3}$ $M_\odot$, whereas a 1.3-$M_\odot$ white dwarf need only accrete $\approx 3\times 10^{-5}$ $M_\odot$ before erupting. At a typical mass-transfer rate of $\sim 10^{-9}$ $M_\odot$ yr^{-1} a cataclysmic variable with a 1.3-$M_\odot$ white dwarf would thus erupt every $\sim 30\,000$ yrs, whereas one with a 0.6-$M_\odot$ white dwarf would erupt every ~ 5 million years. Thus every cataclysmic variable goes nova many times in its $\sim 10^8$–10^{11}-yr lifetime.

It is obviously important to know whether the mass ejected in a nova explosion is greater or less than the amount accreted between eruptions, since this affects the evolution of the system. Estimates of the mass in observed shells are typically 10^{-4} $M_\odot$, compatible with the expected mass of the accreted layer. However, observations of emission lines indicate that the material in nova shells is overabundant in heavy nuclei such as carbon, nitrogen, oxygen and neon, compared to that expected for material transfered from the secondary star. This suggests that the accreted layer can mix with the outer layers of the white dwarf, and that a nova explosion takes with it some of the white dwarf's matter. Thus white dwarfs in cataclysmic variables lose matter over time, rather than gaining it. There may, though, be an exception for the heaviest white dwarfs when they accrete at high mass-transfer rates; these may gain mass overall.[3]

Since the nature of long-term variations in mass-transfer rate is poorly understood (Section 12.5), there has been much speculation about the effect of nova eruptions on mass transfer. Among the suggested effects are (1) the loss of material causes the binary separation to increase; (2) the friction in the common-envelope phase of the eruption causes the binary separation to decrease, (3) the irradiation of the secondary by the nova increases the mass-transfer for a period; once it has cooled the secondary therefore underfills the Roche lobe and so drops to a much lower mass-transfer rate. Observationally there appears to be no systematic difference in the mass-transfer rates of cataclysmic variables observed before and after

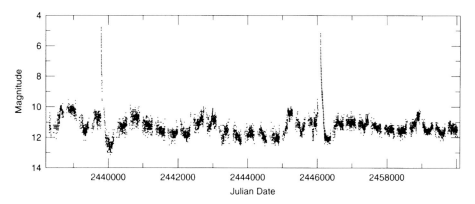

Fig. 11.6: The lightcurve of RS Oph over 32 years, showing two outbursts (in Nov 1967 and Jan 1985) and slower variability during quiescence. The data are 1-day averages compiled by the AAVSO.

nova eruptions, which suggests that the effects above are minor or short-lived. Indeed, some old novae are seen to undergo dwarf-nova outbursts whereas others have stable discs, indicating that there is a spread in mass-transfer rates decades after a nova. It should be noted, however, that our observational record is very short compared to nova recurrence times, so we may simply be looking at short-term fluctuations in a longer-term pattern.

11.4 RECURRENT NOVAE

If a system is seen to undergo more than one nova eruption it is allocated to the sub-class *recurrent nova* (whereas the term 'classical nova' is applied to novae which have not been seen to repeat). Are recurrent and classical novae the same objects, differing only in recurrence timescale? Some of them might be: for systems with white-dwarf masses of 1.4 $M_\odot$ accreting at rates of $\gtrsim 10^{-8}$ $M_\odot$ yr^{-1}, the recurrence time drops to < 100 yrs. T Pyx is the best candidate to be such a system.

However, most other recurrent novae are not typical cataclysmic variables but have very long orbital periods (days or hundreds of days) and secondary stars that are giants or are evolving to become giants. A good example of such a system is RS Oph, a 230-day binary with a red-giant secondary whose lightcurve is shown in Fig. 11.6. The nature of the outbursts in such systems is unclear, and while many may be thermonuclear runaways on the white dwarf, some of the events classed as recurrent-nova eruptions could be accretion events (either from Roche-lobe overflow or by wind accretion), or disc instabilities.

11.4.1 Relation to supernovae?

There is one event yet more dramatic than the nova explosion — the supernova explosion. A supernova is the explosive disintegration of an entire star, and one type of supernova, called 'Type 1a', is thought to occur when a white dwarf is pushed over the Chandrasekhar limit of 1.4 $M_\odot$. If the white dwarf is composed mostly of carbon and oxygen, the most likely outcome is the detonation of nuclear burning of these elements, releasing sufficient energy to blow the star apart. Alternatively, if the white dwarf is rich in iron — which being the most stable element cannot undergo further burning — the white dwarf would collapse into a 'neutron star'. This occurs when the degeneracy pressure on the electrons becomes so high that the electrons are forced into the nuclei, where they combine with the protons to create neutrons. The disappearance of electrons removes the degeneracy pressure; with nothing to support the star it collapses to a radius of ~ 10 km, at which point the neutrons become degenerate and their pressure prevents further collapse. The neutron star is essentially a giant atomic nucleus, 10 km across.

Such supernovae are rare — the last one in our Galaxy appeared 400 years ago — but the disintegration of a white dwarf releases so much energy that the resulting supernova can outshine the combined output of an entire galaxy.

Does the accretion of material onto the white dwarf in a cataclysmic variable ultimately lead to a supernova? As outlined above, current observations suggest that most white dwarfs lose more mass through nova eruptions than they accrete, but the white dwarfs in recurrent novae may gain mass, and thus might eventually go supernova. Thus, although there is yet no proof, recurrent novae, along with mergers of binary white dwarfs, and the 'supersoft' sources discussed below, are candidates for being the progenitors of Type 1a supernova explosions.

11.5 BURSTS IN NEUTRON-STAR SYSTEMS

The formation of neutron stars as described above can lead to 'cousins' of cataclysmic variables, in which a red dwarf transfers matter onto a neutron star, rather than a white dwarf (these are called *low-mass X-ray binaries*, see Section 13.1).

As with white dwarfs, the accreting material can undergo a thermonuclear runaway, and produce the counterpart of a nova eruption. However, the much higher gravity and pressure on the surface of a neutron star (with 1-$M_\odot$ of material condensed into a ~ 10-km radius, rather than ~ 7000 km) leads to a runaway after the accretion of only 10^{-12} $M_\odot$ of material. This brings the recurrence time down to only hours. Such events are referred to as *bursts*.[4] Fig. 11.7 shows two bursts from the low-mass X-ray binary EXO 0748–676, which last less than a minute and occur only 3 hrs apart.

The neutron stars in X-ray binaries are spun up by accretion so that their spin periods are typically only milliseconds. During the rise of the burst, we sometimes see a strong pulsation at this period (see Fig. 11.8). We deduce that the thermonuclear runaway began at one location on the neutron star's surface, producing a bright spot that comes into and out of view as the star spins. The thermonuclear runaway takes 1–2 secs to spread over star, and then the pulsation disappears. Presumably a similar effect might occur during the ignition of an eruption on a white dwarf, but we have never observed a nova early enough to detect it.

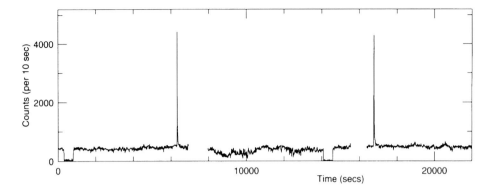

Fig. 11.7: This X-ray lightcurve of EXO 0748–676 shows two 'bursts' caused by thermonuclear runaways in accreted material on the surface of a neutron star. Note also the eclipses and dipping activity, seen more easily at the scale of Fig. 7.7. (Data from the *EXOSAT* satellite.)

11.6 SUPER-SOFT SOURCES

As the mass-accretion rate increases, the interval between nova eruptions decreases. If $\dot{M}$ were increased to $\approx 10^{-7}$ $M_\odot$ yr^{-1} (far above the usual values of 10^{-10}–10^{-8} $M_\odot$ yr^{-1}) burning of hydrogen to helium would occur continuously, rather than episodically. If $\dot{M}$ were increased further, to $\gtrsim 4 \times 10^{-7}$ $M_\odot$ yr^{-1}, the inflow of material could not be assimilated; it would form a cocoon around the white dwarf, turning it into a red-giant star.

Mass transfer within the narrow range allowing continuous nuclear burning is thought to give rise to a class of objects called *super-soft X-ray sources*.[5] The

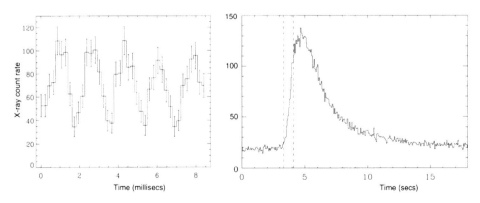

Fig. 11.8: *Right:* A burst from the X-ray binary X1636–53 recorded with the *RXTE* satellite on the 1998 August 20. When analysed more closely, the data from the burst rise (shown as dashed lines) reveal a prominent pulsation at the 1.7-msec spin period of the neutron star (*left*). (Figures courtesy of Tod Strohmayer.)

high $\dot{M}$ and nuclear burning make such stars 10 000 times more luminous than normal cataclysmic variables, giving them an energy output comparable to that from accretion onto a neutron star. The energy from nuclear burning heats the white-dwarf surface to $>100\,000$ K, so that it emits intense blackbody emission peaking at soft-X-ray energies of ≈ 50 eV. This contrasts with the harder 1–10-keV emission from an accreting neutron star, and is the reason for the 'super-soft' label.

11.6.1 Thermal-timescale mass transfer

Super-soft X-ray sources are much rarer than cataclysmic variables, since the high $\dot{M}$ needed to drive them occurs only under particular evolutionary conditions, and then lasts for a shorter time. To understand the high $\dot{M}$ we need to return to the discussion of mass transfer in Section 4.2. Recall that if the mass-donor is more massive than the mass-gaining star then the transfer of material causes the Roche lobe of the donor star to shrink, and, since this precipitates further mass transfer, a runaway results. Thus mass-transfer is only stable when the secondary is lighter than the white dwarf.[†]

However, a more careful analysis needs to consider the response of the secondary star to the loss of mass.[6] If the secondary tends to expand after losing its outer layer (as occurs in some stars with strong convective turbulence in their outer envelopes) unstable mass transfer can occur for mass ratios as low as $q \approx 0.7$. Conversely, if the secondary contracts on mass loss, and shrinks to a greater extent than the Roche lobe, mass transfer will be stable. This is most likely in stars with no convective motions in their envelopes, and mass-transfer from such stars is stable up to $q \approx 1.3$.

Understanding super-soft sources involves one further complication, namely that the response of the secondary star to mass loss takes place in two stages. The first stage a rapid adjustment to ensure that pressure forces are still in equilibrium with the gravity holding the star together. The second stage is an adjustment of the star's internal temperature structure to the new situation. This requires the flow of energy, which takes place on a much longer 'thermal' timescale than the 'dynamical' timescale on which pressure equalises.

Now, if the secondary's response to mass loss is that it contracts on the dynamical timescale, but expands again on the thermal timescale, then mass-transfer will still be unstable, but instead of proceeding by a runaway, it will be slowed down to the thermal timescale on which the secondary expands. This *thermal-timescale mass transfer* is much larger than that usually seen in cataclysmic variables and can be sufficient to turn the system into a super-soft source.

It is possible that many observed super-soft sources are recently created cataclysmic variables with mass ratios greater than ≈ 1; they would undergo a period of thermal-timescale mass transfer continuing until the mass ratio had decreased to less than ≈ 1, whereupon they would settle down to mass transfer at the lower rates driven by magnetic braking.

[†]The analysis of Box 4.1 shows that the secondary's Roche lobe will decrease on mass transfer if $q > 5/6$. Thus stability requires $q < 5/6$.

Chapter 12

Secondary star variations

Most of this book has focused on the accretion flow — the stream, the disc, the impact with the white dwarf — which grabs the attention by emitting most of the radiation and producing most of the variability of a cataclysmic variable. The secondary star has been largely relegated to a mere supplier of an accretion stream.

However, it is increasingly realised that variations in the mass-transfer rate, $\dot{M}$, take place on timescales of days, months, years and upwards, although the reasons for this remain unclear and are the source of much debate.

A naive application of evolutionary theory (Chapter 4) would suggest that $\dot{M}$ is determined solely by the orbital period and the masses of the two stars. But as has been pointed out in earlier chapters (Sections 4.2.2 and 5.4), systems that are very similar in these respects can have $\dot{M}$ values differing by factors of 100–1000. Thus to complete the survey of cataclysmic variability this chapter asks what happens when the accretion rate changes.

12.1 VY SCL STARS

VY Scl stars can be classed as novalike variables, spending much of their time with high mass-transfer rates and hot discs. However, they also drop by several magnitudes into *low states* during which $\dot{M}$ drops or shuts off completely (see Fig. 12.1).

VY Scl stars have sometimes been referred to as 'anti-dwarf-novae' with the implication that their behaviour is the opposite to that of dwarf novae, and might also be caused by a disc instability. However, the lightcurves and timescales are very different from those of dwarf novae: while Z Cam stars also spend long periods in a high state, the rest of the time they show a rapid succession of outbursts which have rise and decline times of days. When a VY Scl star drops out of a high state it shows little or no dwarf nova activity, implying that the mass-transfer rate must have dropped to well below that of a Z Cam (compare Fig. 12.1 with Fig. 5.17). This is supported by the amplitude of the VY Scl variation, which is much larger than that of a Z Cam (typically 4.5 mags compared to 2.5 mags). Further, the low states of VY Scl stars appear random, with none of the quasi-periodic behaviour associated with dwarf novae.

172   Secondary star variations   Ch 12

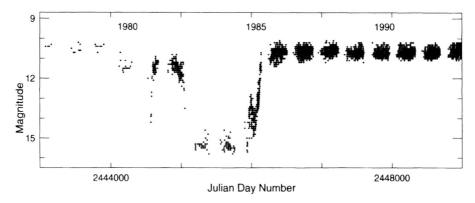

Fig. 12.1: The lightcurve of TT Ari, a member of the VY Scl subclass, showing deep low states in an otherwise-stable novalike lightcurve. (Data compilation by the AAVSO, AFOEV and VSOLJ.)

Clinching evidence that VY Scl low states are not related to discs comes from their occurrence in the AM Her stars (see Fig. 12.3). Since AM Her stars do not have discs, there is no 'reservoir' to smooth the accretion flow, so all variations longer than the orbital cycle must be changes in $\dot{M}$.

This does, though, raise a puzzle. If the mass-transfer sustaining the disc in a high state is suddenly switched off, the disc would drop out of the high state, but

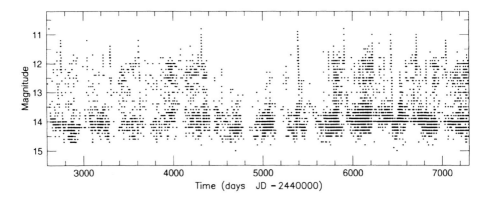

Fig. 12.2: In dwarf novae, changes in the mass-transfer rate are hard to detect, since the disc acts as an unstable reservoir, depositing accretion onto the white dwarf episodically. A reduction in $\dot{M}$, however, will lead to a reduction in the frequency of dwarf-nova outbursts. On the scale of this 12-yr plot of SU UMa, the outbursts are too rapid to discern individually, but it is noticeable that for a ∼3-yr period the number of outbursts was markedly lower, which presumably was the equivalent of a VY Scl low state. The plot is based on work by Rosenzweig and colleagues,[1] using data from the AAVSO.

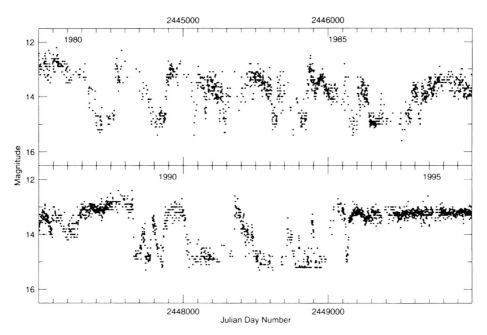

Fig. 12.3: 16 years in the life of AM Her, showing frequent, erratic drops into low states. Since AM Her does not have a disc, this plot can also be considered as a plot of $\dot{M}$ against time. The data are 1-day averages compiled by the AAVSO.

then undergo a series of dwarf-nova outbursts triggered by material diffusing into the inner disc. Since only $\sim 10\%$ of a disc's mass is removed by each outburst, many outbursts could occur before the disc could no longer trigger outbursts. However, such outbursts are not seen. One suggestion is that in VY Scl stars the white dwarfs are much hotter than in dwarf novae, heated by the higher average $\dot{M}$. The hot white dwarf could irradiate the inner disc, ensuring that it remained in a hot, viscous state, keeping the accretion rate high, and draining the disc of most of its mass before it dropped out of the high state.[2] It is worth noting, though, that the low states of most VY Scl stars are poorly studied, since they are relatively faint, and outbursts may have been missed. The spread of CCD cameras amongst amateurs should start to redress this.

The reasons for the reductions in $\dot{M}$ that cause VY Scl low states are uncertain, though most explanations invoke magnetic activity on the secondary star. For instance, star spots — the counterpart of the spots seen on our Sun, which are areas of lower temperature caused by magnetic activity — might pass over the Lagrangian point. Since spots are slight indentations in a star's surface, this would reduce or end mass transfer until the spot had moved on.[3]

This is only one of a number of indications that secondary stars in cataclysmic variables have strong magnetic activity. As already discussed in Chapter 4, a magnetic field on the secondary star is needed to drive mass transfer, while the locking of the orbital and spin periods in AM Her stars suggests a coupling between the fields of the primary and secondary.

12.2 OBSERVING STAR SPOTS

So do secondary stars show spots? Obviously we don't have the resolution to image a secondary star and see spots directly, but a method suggested by Tim Naylor provides strong evidence for their existence. The method uses the fact that a spot has a lower temperature than its surroundings, and thus has a different spectrum. A spotty secondary will thus have a spectrum that differs from a non-spotty red dwarf of the same temperature, but matches a composite of the spectrum of two red dwarfs, one cooler than the other. This is exactly what was found for the spectrum of SS Cyg (Fig. 12.4): a composite spectrum matches better than any single-temperature spectrum, indicating that 22% of the red dwarf in SS Cyg is covered in spots, and thus that it is magnetically active.

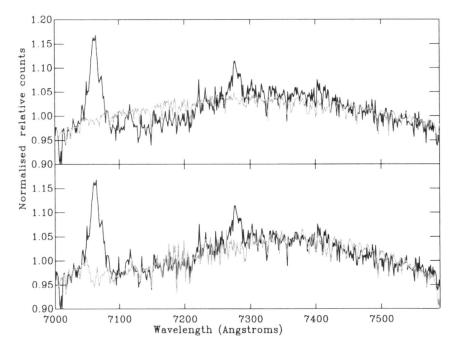

Fig. 12.4: The spectrum of SS Cyg is shown in black in both panels. In the upper panel the grey line is the best-fitting model using a single-temperature red-dwarf spectrum. The main mismatch is at 7100–7200 Å, where TiO features are strong in cooler red dwarfs. In the lower panel the grey line is the best-fitting model composed of red-dwarf spectra at two temperatures, to simulate a spotty secondary. The fit is much better. The emission features at 7060Å, 7275Å and 7410Å, from the accretion disc, were not included in the fitting. (Figure by Natalie Webb and Tim Naylor.[4])

12.3 SOLAR-TYPE CYCLES

Our Sun is an example of a magnetically active star: the combination of convective churning of the Sun's outer layers and the Sun's rotation are thought to drive a dynamo, creating a magnetic field that results in prominences — material held by field lines far above the Sun's surface — and in sunspots where the loops of field lines pass through the visible surface. The magnetic activity varies over an 11-yr cycle, with periods of high sunspot activity alternating with times when few or no sunspots are seen.

If secondary stars in cataclysmics are also magnetically active — and with their 2–10-hr rotation periods being much faster than the Sun's 26-day rotation the dynamo could be far more powerful — we might expect their activity to undergo similar cycles. Indeed, many close binaries exhibit subtle periodicities on timescales of 5–50 yrs that can be attributed to this.[5]

Such periodicities are most readily seen in eclipse timings. If the orbital period is exactly constant, eclipses will occur at exactly predictable times. But sometimes the eclipses occur earlier or later than expected, and can vary cyclically between the two. Since such cycles tend to be quasi-periodic, rather than repeating exactly, they cannot be explained by the gravitational influence of a third star in the system, and instead are probably solar-type cycles.

For example, the eclipses of U Gem wander about the expected times by ~ 1 min on a timescale of ~ 8 years. This translates to a change in the orbital period of one part in a million (the difference then accumulates each cycle to produce a measurable effect). The favoured explanation for this behaviour is that the cyclic change in

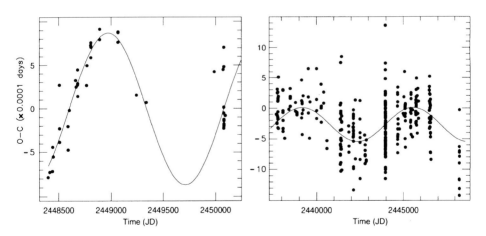

Fig. 12.5: The wanderings of eclipse times of EX Dra (*left*) and EX Hya (*right*) are revealed by plots of 'observed' minus 'calculated' times ($O - C$ values). EX Dra exhibits an apparent ~ 4-yr variation while EX Hya exhibits a ~ 17-yr variation (both traced by illustrative sinusoids). Since the data span is not much greater than the suggested periods, however, these results are uncertain. The EX Dra plot is based on work by Raymundo Baptista and Hauke Fiedler.[6]

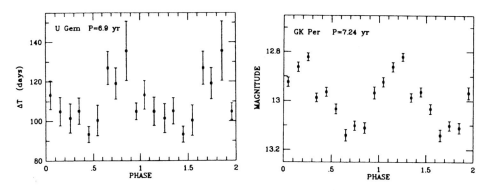

Fig. 12.6: *Left:* the interval between outbursts of U Gem, varying over a 7-yr cycle. *Right:* the quiescent magnitude of GK Per, varying over a 7-yr cycle. (Figures by Antonio Bianchini.[7])

magnetic activity alters the distribution of the angular momentum associated with the rotation of the secondary star, and hence changes its shape slightly. The change in shape affects the gravitational attraction between the two stars in the binary and this, together with the fact that orbital angular momentum is conserved, changes the orbital period.[8]

The change in orbital period can affect the mass-transfer rate. Thus magnetic cycles are seen as slow waves in the brightness of novalikes, or in the brightness of dwarf novae when at minimum light. An increase in $\dot{M}$ in a dwarf nova will also trigger more frequent outbursts, and it is notable that the interval between outbursts in U Gem shows a variation on a timescale of ~ 7 years, compatible with the timescale of its eclipse wanderings (see Fig. 12.6).

12.4 SHORT-LIVED FLARES: BURSTS OF MASS TRANSFER?

While it is widely accepted that dwarf-nova outbursts result from unstable accretion discs, there have been many observations of shorter-lived 'flares' whose origin is less clear. These are brightenings of ~ 1–2 magnitudes, lasting only for hours, whereas dwarf-nova outbursts last for days. They have occurred both in novalikes and in dwarf novae, where they are often overlooked amongst the outbursts.

Could these result from activity on the secondary star, in some unknown instability that produces a short burst of enhanced mass transfer? There is strong evidence that this does occur (though we don't know enough about discs to rule out a disc origin). For instance, TV Col has shown several flares (one is shown in Fig. 12.7) despite the fact that it is a novalike, with a disc stuck in a stable high state complete with permanent superhumps. The amplitude and duration of this flare also distinguish it from a dwarf-nova outburst. Furthermore, analysis of spectra during a flare showed that the line emission from the bright spot increased thirty-fold, suggesting markedly increased mass transfer. One counterpoint, though, is that TV Col has a magnetic white dwarf, and this would affect a disc-instability

Sec 12.4   Short-lived flares: bursts of mass transfer?   177

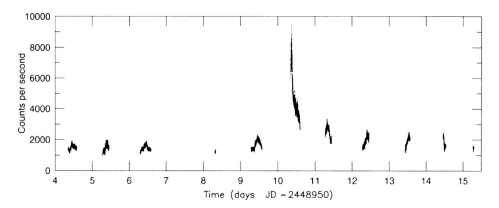

Fig. 12.7: A lightcurve of TV Col showing a brief 'flare' lasting for ~ 12 hrs. Lower-amplitude variations are seen within each night's data.[9]

outburst, possibly allowing it to produce such features.

QS Tel, however, is an AM Her star, with no accretion disc. While it was in a low state (with little or no mass transfer occurring) it produced two flares (Fig. 12.8). They lasted only ~ 1 hr, during which time the ultraviolet light increased by a factor ~ 7. With no disc involved, this can only be attributed to activity on the secondary star. It could either be a mass-transfer event, or possibly magnetic activity on the secondary star generating the ultraviolet light directly (similar, though far weaker, 'solar flares' are seen on our Sun). Thus to summarise, although it is hard to

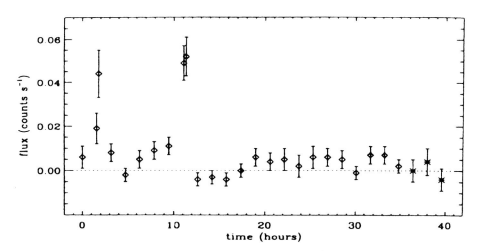

Fig. 12.8: The ultraviolet lightcurve of the AM Her star QS Tel, showing two 'flares' upon a much lower quiescent level. The data, covering the 65–190Å range, were obtained with the $EUVE$ satellite on 1992 July 8. (Figure by John Warren and colleagues.[10])

12.5 LONG-TERM VARIATIONS IN MASS-TRANSFER RATE

Chapter 4 outlined the basic theory of cataclysmic variable evolution. Above the period gap, the magnetic field of the secondary star, generated by fast rotation and convective turbulence, couples with a wind to drive mass transfer by magnetic braking. Below the gap, gravitational radiation drives mass transfer. With the secondary-star mass, radius and rotation rate all determined by the orbital period (see Appendix A), the result should be a set mass-transfer rate for each orbital period (with possibly a weak dependence on the the mass ratio q). Yet at some orbital periods we observe novalikes (high-$\dot{M}$ systems), and dwarf novae (low-$\dot{M}$ systems), and the borderline Z Cam stars, spanning a range of $\dot{M}$ of at least 100. Furthermore, there are systems below the gap with $\dot{M}$ well above that explainable by gravitational radiation. This is illustrated by Fig. 12.9.

Clearly, some additional factor causes variations in $\dot{M}$, and acts on a longer timescale than the monthly or yearly variations attributable to magnetic activity of the secondary star. Yet on the longest timescales, the average $\dot{M}$ predicted by the evolutionary arguments should surely hold.

The nature of the mechanism causing long-term deviations from the evolutionary $\dot{M}$ is currently being debated, without yet being settled. One relevant factor might be the effect of nova explosions. For instance, the loss of material during an eruption would cause the binary separation to increase. Counteracting this is the friction during the common-envelope phase, causing the separation to decrease. The overall change thus depends on the speed of the nova, and hence the duration of the common-envelope phase, but is thought to be small and not significant in the long term. A third effect is the greater irradiation of the secondary by the hot white dwarf in the decades after the nova eruption. One proposal is that this increases $\dot{M}$ above the secular mean, and so drives the secondary out of equilibrium. Once the irradiation has died down the secondary would relax, underfilling its Roche lobe, so that $\dot{M}$ decreases to near zero. The system would be too faint to be recognised as a cataclysmic variable, a state called 'hibernation'. Eventually, evolution would drive the system back into contact, and mass transfer would resume.[11]

This offers the possibility of cycling between subtypes — from high-$\dot{M}$ to low-$\dot{M}$ and back, perhaps appearing as a novalike, then a dwarf nova, then hibernating, and then returning to dwarf-nova or novalike status before undergoing another nova eruption. The difficulty with such ideas is their lack of observational support, in that no old nova has been observed to go into hibernation. However, we have only a limited length of record and a limited number of systems upon which this can be tested, so the hypothesis is not yet ruled out.

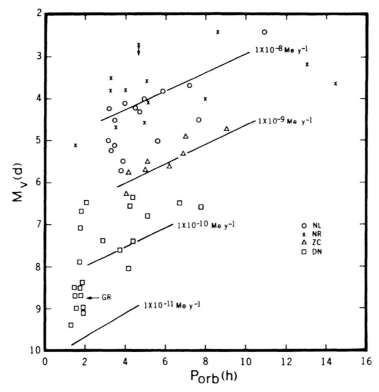

Fig. 12.9: This compilation, by Brian Warner,[12] shows the brightness (absolute magnitude) of the disc plotted against orbital period. The corresponding mass-transfer rates are illustrated by lines (which slope upwards since discs are larger for higher P_{orb} and so are brighter for the same $\dot{M}$). The different symbols denote novalikes (NL), old novae (NR), Z Cam stars (ZC) and dwarf novae (DN). The 'GR' arrow shows the level attributable to gravitational radiation.

12.5.1 Irradiation-induced mass-transfer cycles

Regardless of the effect of nova eruptions, it is likely that irradiation of the secondary by the accretion luminosity plays a major role in determining $\dot{M}$. The basic principle is that a feedback loop can form, in which higher irradiation leads to greater $\dot{M}$, which leads to a hotter white dwarf and boundary layer, leading to higher irradiation, and so on.[13] Eventually the increase is stopped because higher $\dot{M}$ also leads to increased flaring of the disc, blocking the irradiation of the secondary. The converse of the above is that a low-$\dot{M}$ state would produce little irradiation and thus retain a low $\dot{M}$. Overall, a state with $\dot{M}$ far above the evolutionary mean could be self-sustaining until it drove the secondary too far out of equilibrium, and $\dot{M}$ dropped. As the white dwarf cooled, the system would experience a period of much lower $\dot{M}$, until angular-momentum loss had decreased the Roche lobe sufficiently to restore a higher $\dot{M}$ and initiate another cycle.

There is some circumstantial evidence for this effect. Firstly, there are relatively few dwarf novae with periods just above the period gap ($P_{orb} \approx 3$–4 hrs) compared to longer periods (Fig. 5.16), suggesting that in this period range few systems undergo mass transfer at medium rates. This could be because, of the systems above the period gap, those with the shortest periods have the smallest separations, where irradiation effects will be most intense.* Systems below the gap have even smaller separations, but here the lower starting $\dot{M}$ may be insufficient to trigger a feed-back cycle. Furthermore, all of the VY Scl stars with securely established orbital periods are in the 3–4-hr range. Again, this suggests that such systems cannot exist with moderate $\dot{M}$s, but are forced to cycle between states of very high and very low $\dot{M}$. The SW Sex stars are also preferentially found in this period range, and, as suggested in Section 7.4.5, some of their peculiarities may result from extreme $\dot{M}$, compensated for by time spent in VY Scl low states.

To summarise: although there are difficulties in proving any model for variations longer than our observational record, irradiation-induced mass-transfer cycles are the most plausible mechanism for the long-term departures of $\dot{M}$ from the evolutionary average, and may explain why several cataclysmic variables with the same orbital period can exhibit very different behaviour.

---

*Irradiation falls off by the inverse-square law, and so will decrease by a factor 3 as P_{orb} is increased from 3 to 6 hrs.

Chapter 13

Variations on the theme

By definition, cataclysmic variables contain white dwarfs that accrete material pulled from a close companion. But such systems constitute only one type amongst the wide variety of interacting binary stars. Resisting the temptation to address each type at length, this chapter aims only to highlight the fact that many of the phenomena occurring in cataclysmic variables have parallels in other astronomical objects. In fact, accretion is one of the most widespread and important processes in astrophysics.

Most binaries are in wide orbits, with little or no interaction between the stars. As one moves to shorter orbits, the probability of seeing eclipses increases, since the requirement that they be seen edge-on becomes less stringent. Stars in closer binaries can also induce tides in their companion, spinning them up by locking their rotation to the orbit, and thus increasing their magnetic activity. RS CVn stars fall into this category. Other ways of influencing a companion include irradiation and stellar winds.

In even closer binaries, the stellar sizes become comparable to the Roche lobes, allowing the direct exchange of matter. When both stars fill their Roche lobes they are in 'contact', forming a common envelope around two cores (see also Box 2.3). The common envelope in such 'W UMa stars' has a uniform surface temperature, and thus the orbital modulation is caused simply by the change in projected surface area as the 'dumbbell' rotates. This produces a characteristic lightcurve with rounded maxima and unequal minima (Fig. 13.1).

The 'semi-detached' binaries — with one star in contact with its Roche lobe — include the cataclysmic variables, and also systems in which the mass-gaining star is not a white dwarf. In 'Algol' systems, named after the 2nd-magnitude star in Perseus, a normal star accretes matter from a lower-mass companion, which has started evolving towards becoming a red giant and so has filled its Roche lobe. This sounds paradoxical, since in theory the higher-mass star should evolve faster, and so become a red giant first. The solution is that the evolved star was initially the heavier. Thus when it first filled its Roche lobe it initiated mass transfer from a heavier star to a lighter one, which is unstable (see Section 4.1). The ensuing mass transfer would have rapidly depleted the star, reversing the mass ratio, until the

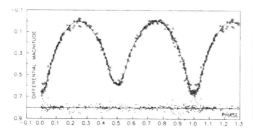 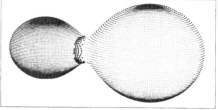

Fig. 13.1: The B-band lightcurve of the W UMa system RZ Tau, showing the characteristic lightcurve of a rotating dumbbell. At right is an illustration of the system at phase 0.7. (Figure by Djurašević, Zakirov and Erkapić.[1])

binary settled down to the phase of more leisurely mass transfer that we see as an Algol system.

13.1 X-RAY BINARIES

The pressure of degenerate electrons can only support a white dwarf up to a limit of ≈ 1.4 M$_\odot$. Above this mass the intense pressure forces the electrons to combine with protons to produce neutrons. The absence of electrons removes the electron-degeneracy pressure, and the star collapses until stopped by neutron-degeneracy pressure, at a radius of only 10 km (cf. the 7000-km radius of a white dwarf). Such an object is called a neutron star.

With a much smaller radius, but a similar mass to a white dwarf, the gravitational field at the surface of a neutron star is 500 000 times as intense. The energy

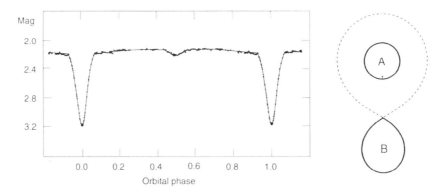

Fig. 13.2: The lightcurve of Algol in Hα light, together with an illustration of the system. Star A is hotter and more massive (12 500 K and 3.7 M$_\odot$) while Star B is cooler and less massive star (4500 K and 0.81 M$_\odot$); it is evolving to become a red giant and so fills its Roche lobe. Note the primary eclipse (of A by B), the secondary eclipse (of B by A) and a reflection effect caused by the irradiation of B by A. (Figure adapted from work by Richards, Mochnacki and Bolton.[2])

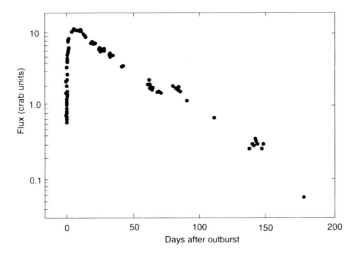

Fig. 13.3: The X-ray lightcurve of the soft X-ray transient GS2000+25, during its 1988 outburst. The lightcurve shows the typical outburst profile of an SXT, with a fast rise and an exponential decay on the disc's viscous timescale. The data, obtained with the *Ginga* satellite, cover the 1–20-keV band and are in units of the flux of the Crab nebula.

released by accretion onto a neutron star is thus increased 700-fold, and most of this emerges as X-rays. The 'X-ray binaries', in which matter accretes onto a neutron star, are the brightest X-ray sources in the sky, even though they are much rarer than cataclysmic variables, and so typically 10 times more distant.[3]

One class of X-ray binary, the low-mass X-ray binaries (or LMXBs), are sufficiently similar to cataclysmic variables that we have already encountered them in Sections 7.3 and 11.5. They again accrete material pulled from a red-dwarf companion, and can be regarded as cataclysmic variables which harbour a neutron star rather than a white dwarf (though some contain, instead, a yet more massive and more compact black hole).

The fierce irradiation in an LMXB means that the energy of the accretion disc is not dominated by viscous dissipation, as in a cataclysmic variable, but by the absorption of X-rays and their re-radiation in the UV and optical. One consequence of this is that systems containing neutron stars rarely show the equivalent of dwarf nova outbursts, since the irradiation ensures that the disc is always hot and ionised.

13.1.1 Soft X-ray transients

Some LMXBs, however, do show outbursts, with X-ray fluxes rising by factors of ~ 1000 for periods of months (see Fig. 13.3). Such systems are called 'soft X-ray transients', or SXTs, because of their soft X-ray spectrum.*

---

*An alternative name is 'X-ray novae', but this is misleading because the behaviour is more akin to that of dwarf novae than novae.

When the masses of SXTs are derived it is generally found that the compact star is typically 5–15 $M_\odot$, greater than the 3-$M_\odot$ limit for the mass of a neutron star. Above 3 $M_\odot$ a compact star will collapse into a black hole, possessing a gravitational field so strong that not even light can escape.

Since the 'Schwarzschild radius' of a 10-$M_\odot$ black hole — the point of no return on approaching it — is comparable to the 10-km radius of a neutron star, it can be difficult to determine whether a particular binary contains a neutron star or a black hole. However, the major difference is that a neutron star has a solid surface, and accreting material slamming into it gives rise to hard-X-ray emission. Furthermore, the accreted material can then undergo thermonuclear bursts. In contrast, a black hole has no surface, so accreting material just disappears, taking with it much of the energy. We see only the strong UV and soft-X-ray emission from the inner disc surrounding the black hole. However, the division into systems with a soft X-ray spectrum (implying a black hole) and those with a hard X-ray spectrum (implying a neutron star) is imperfect, since at low accretion rates the inner disc of both types of system can evaporate into a coronal flow, which emits hard X-rays.

Since the inner disc is less effective than a neutron star at irradiating the outer disc (due to the greater foreshortening) the outer disc in a black-hole system can exist in a cool, un-ionised state. Thus, such discs can undergo the thermal instability and produce dwarf-nova-like outbursts. This explains why SXTs are generally found to be black-hole systems.[4]

The outbursts in SXTs are, however, lengthened by irradiation. In dwarf novae the outburst ends when a cooling wave moves through the disc in 1–2 days (the 'thermal timescale' on which disc material can heat or cool the next annulus). In SXTs the irradiation prevents the cooling wave from moving inwards, and so the disc is maintained in a high state. It is gradually depleted of material on a longer timescale of ~ 1 month (the 'viscous timescale' on which material flows through the disc). Thus SXT discs are nearly completely accreted by each outburst (whereas only $\sim 10\%$ of a dwarf-nova disc is accreted during an outburst), and so decades can elapse before the disc is sufficiently replenished for the next outburst (compared to the $\sim$ monthly recurrence of dwarf novae).

13.1.2 High-mass X-ray binaries

In low-mass X-ray binaries the mass-donating star is less massive than the mass-gaining neutron star or black hole — only then is mass transfer by Roche-lobe overflow stable. However, a separate class of 'high-mass X-ray binaries' involves mass donors which are much more massive than the accreting object. These instead transfer matter by a 'stellar wind'.

Stars with masses in the range 10–30 $M_\odot$ are thousands of times more luminous than the Sun. At such luminosities the pressure of radiation on material in the outer layers can generate a strong outflow of particles, in which as much as 10^{-6} $M_\odot$ yr^{-1} of material is driven off in a stellar wind (in contrast the 'solar wind' of our Sun amounts to only 10^{-14} $M_\odot$ yr^{-1}). Some small fraction of this outflow can be captured by the gravity of an orbiting neutron star, to spiral around the star

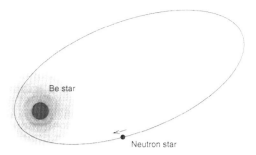

Fig. 13.4: A schematic plan view of a Be–neutron-star binary. The Be ejects a disc of material in its equatorial plane. During most of its elliptical orbit the neutron star is too far from the Be star to accrete this material, but near closest approach it becomes a strong X-ray source.

and eventually accrete.

A further method of mass transfer occurs in Be–neutron-star binaries. The Be stars are again massive stars (~ 10 M$_\odot$) but are rotating so rapidly that they are throwing off material from their equator. This material, expelled by centrifugal forces, forms a ring of material around the Be star. An orbiting neutron star passing through the ring will accrete some of this material. In many of these binaries the neutron star is in an elliptical orbit (normally, in close binaries, tidal forces circularise the orbit of the lower-mass star, but when this star is as small as a neutron star, tidal forces are negligible). In such systems the neutron star might pass into the ring of material only near closest approach to the Be star (see Fig. 13.4). These systems are transient X-ray sources, since no accretion or X-ray emission occurs for the majority of the orbit.

One feature of the high-mass X-ray binaries is that most of their neutron stars are highly magnetic, having surface fields of typically 10^{12} G. Thus, the X-ray emission is often pulsed at the neutron-star spin period, which is typically in the range 1–100 secs.

13.1.3 The micro-quasars

Take a look at the X-ray lightcurves in Fig. 13.5, all resulting from accretion onto a black hole. One's first reaction is surprise that different systems can produce such different lightcurves. But surprise turns to astonishment when one realises that these are all from the same system, GRS 1915+105, as observed on different occasions. In a book which is essentially a celebration of the lightcurves of accreting binaries, it is fitting that the last binary discussed has the most bizarre behaviour of all.

The lightcurves of GRS 1915+105 are not understood, but the following issues are thought to be important. First, GRS 1915+105 is probably accreting at a rate

186 Variations on the theme   Ch 13

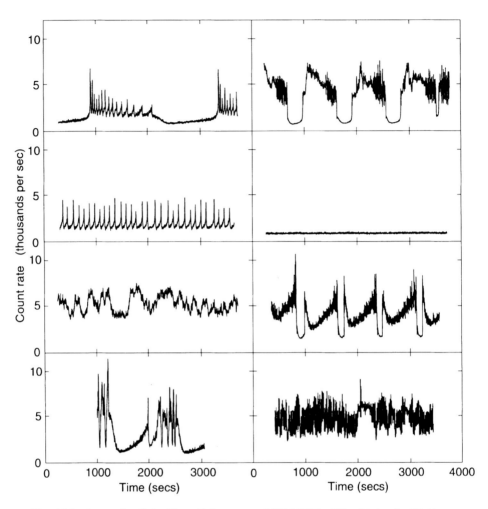

Fig. 13.5: A sample of the X-ray lightcurves of GRS 1915+105, obtained with the RXTE satellite in the 2–30-keV band. All the plots are on the same scale, and illustrate the astonishing range of behaviour this one star exhibits. (Based on work by Michael Muno, Edward Morgan & Ronald Remillard.[5])

close to its 'Eddington limit'. If a system attempts to accrete at a rate above its Eddington limit, the resulting radiation generates enough outward pressure to prevent the accretion of more material. Thus accretion in GRS 1915+105 is probably unstable, occurring in spurts as the attempted accretion rate hovers around this threshold.

Secondly, one way in which GRS 1915+105 releases energy is through the generation of jets of material, which are ejected perpendicular to the accretion disc at speeds which are a large fraction of the velocity of light. The jets are thought to be driven by the intense radiation generated in the inner disc around the black hole, and are collimated by magnetic fields generated by a dynamo effect as the inner-disc material swirls into the black hole.

Another way in which trapped energy might dissipate is by evaporating the inner accretion disc. It is thought that in GRS 1915+105 the accretion energy switches unstably between driving jet outflows and evaporating the inner disc. The constant re-formation and evaporation of the inner disc, and the switching to drive a jet outflow, are probably the origin of much of the variability in the X-ray lightcurves.

The jets in GRS 1915+105 propel beams of electrons far out of the binary, where the combination of free electrons and magnetic fields gives rise to synchrotron radiation that can be detected with radio telescopes. This has led to such systems being called 'micro-quasars', in analogy with the true quasars to which we now turn.

13.2 ACCRETION OUTWITH BINARIES

The occurrence of accretion is not restricted to binary stars; indeed some of the most important occurrences involve accretion onto a lone star or black hole. For example, most galaxies are thought to harbour a super-massive black hole at their centre, weighing perhaps 10^6–10^8 M$_\odot$. Such black holes build up through the accretion of whole stars which, if they stray too close, are shredded by tidal forces and swallowed. When the central regions are densely populated by stars, the black hole has a continual supply of material to accrete, and is referred to as an 'active galactic nucleus' or AGN.[6]

The material from shredded stars forms an accretion disc around the black hole, but one scaled up by many orders of magnitude compared to those in X-ray binaries, to a radius of hundreds of light years. Again, the inner disc can generate jets of material perpendicular to the disc, which extend outwards for many times the size of the galaxy, creating vast radio-emitting lobes. Such galaxies are called 'radio galaxies'. At cosmological distances, the underlying galaxy may not be visible, and we would see only the light from the accretion disc (as a star-like point source) and the radio lobes. Thus they are called 'quasars', as an abbreviation of 'quasi-stellar radio source'.

It is probable that much of our understanding of accretion discs, gained from studying close binaries, is applicable to the discs in AGN. For instance, AGN discs may undergo dwarf-nova-like thermal instabilities. However, the vast size of the disc means that the timescale for an outburst is thousands of years, and so cannot be easily observed. Perhaps, though, the AGN we see are all in a high state, while those galaxies whose nuclei appear inactive are merely in quiescence.

13.2.1 Discs around young stars

A further example of accretion onto single objects occurs during the formation of stars. As a cloud of interstellar gas and dust collapses into a star, it contracts by a factor of a million. Any initial spin of the cloud will, by conservation of angular momentum, be greatly magnified. Thus the cloud cannot collapse spherically, but

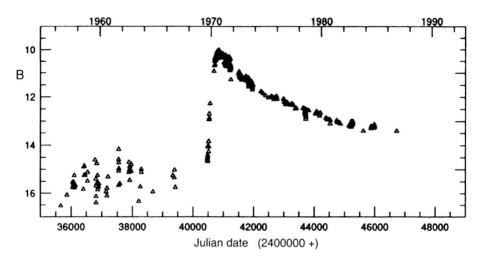

Fig. 13.6: The B-band lightcurve of the FU Ori system V1057 Cyg, showing the equivalent of a dwarf-nova outburst in the disc surrounding a young star. (Figure by Bell et al., based on data compiled by Scott Kenyon and Lee Hartmann.[7])

is flattened by centrifugal forces into an accretion disc. The disc performs the same role as in binaries, transporting most of the matter inwards to accrete into a star, and expelling a small fraction outwards to carry away most of the angular momentum.[8]

This stage of the formation of a low-mass star is called a 'T Tauri' object. Many of the processes familiar from close binaries also occur here. For instance, the rapid rotation of the young T Tauri star can generate a large magnetic field, which disrupts the inner accretion disc and channels the flow onto the magnetic poles, producing an analogue of an intermediate polar. Also, highly collimated jet outflows, perpendicular to the accretion disc, are sometimes seen. Furthermore, T Tauri stars can undergo disc outbursts, most probably by the same mechanism that causes a dwarf-nova outburst. Such stars are referred to as FU Ori objects (see Fig. 13.6).

After most of the disc material has accreted, the newly formed star remains surrounded by the remnants of the disc. This material cannot accrete, since it carries the majority of the angular momentum of the system. Instead it condenses into planets, as particles in the disc collide and stick together, congregating into 'planetesimals' and then planets. Thus it is likely that we owe our existence to an accretion disc, since our home, Earth, condensed out of one.

Appendix A: Deriving the stellar masses

This Appendix discusses how the basic properties of a cataclysmic variable, such as the stellar masses, are derived. For a fuller account consult Warner (1995).

Fig. A.1 illustrates the orbit of masses M_1 and M_2 with a period $P_{\rm orb}$ and a separation $a = a_1 + a_2$ where $a_1 M_1 = a_2 M_2$. As observed at an inclination i, the orbital velocity, K, of either star is

$$K_1 = \frac{2\pi a_1}{P_{\rm orb}} \sin i \quad \text{and} \quad K_2 = \frac{2\pi a_2}{P_{\rm orb}} \sin i.$$

Given Kepler's law (see Box 2.1)

$$a^3 = \frac{G(M_1 + M_2) P_{\rm orb}^2}{4\pi^2}$$

and using

$$a = a_1 \left(\frac{M_1 + M_2}{M_2} \right)$$

we can rearrange to obtain

$$\frac{(M_2 \sin i)^3}{(M_1 + M_2)^2} = \frac{P_{\rm orb} K_1^3}{2\pi G} \quad \text{and} \quad \frac{(M_1 \sin i)^3}{(M_1 + M_2)^2} = \frac{P_{\rm orb} K_2^3}{2\pi G}.$$

The quantities on the right are functions of the observables K and $P_{\rm orb}$ only. The quantities on the left are referred to as the 'mass functions'.

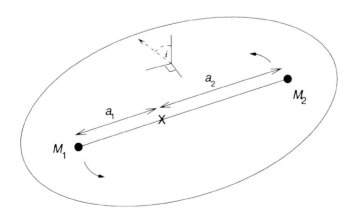

Fig. A.1: Two stars of masses M_1 and M_2 orbit in a plane about the centre of mass, X. The dashed arrow points to Earth, defining the inclination angle i.

If both K_1 and K_2 are known one can use the ratio of the two equations to obtain the mass ratio since

$$\frac{K_1}{K_2} = \frac{M_2}{M_1} \equiv q$$

and from here the individual masses depend only on i. Thus, the standard method of determining the parameters of a binary (assuming that $P_{\rm orb}$ is known) consists of measuring K_1, K_2 and i, or measuring two of these and an additional constraint, such as q, to obtain the third.

A.1 Measuring i and q

The inclination can be reliably measured only in an eclipsing system. The very fact that there is an eclipse of the white dwarf constrains the inclination to $75° \lesssim i < 90°$. A grazing eclipse of a bright spot implies $60° \lesssim i \lesssim 75°$. Furthermore, the length of the eclipse (the time for the white dwarf to travel across a chord of the secondary, see Fig. 2.10) is determined solely by i, q and $P_{\rm orb}$. Other features of the eclipse (for instance the eclipse of a bright spot, given that the stream trajectory depends solely on q) can lead to values for q and i individually.[1]

In non-eclipsing systems, modelling the ellipsoidal variation (Section 2.3.1) can yield a relation between q and i, but this requires that the reflection effect, limb darkening and gravity darkening* have been accounted for.[2]

A.2 Measuring K_1

The white dwarf's absorption lines dominate only in the UV, which is unobservable from the Earth's surface. Thus measuring K_1 directly from the Doppler shift of the lines currently requires the *Hubble Space Telescope*. Since *Hubble* is very oversubscribed, this has so far been done only for U Gem.[3]

The standard technique is therefore to measure K_1 using optical lines from the disc. The reasoning is that if the disc is circular and centred on the white dwarf then the centroid of the disc lines gives the white dwarf velocity. The main difficulty is that the impact of the stream disrupts the symmetry of the disc. However, reasoning further, if the stream disrupts only the outer disc, it affects only the low-velocity part of the line profile, and the high-velocity wings will arise from the undisturbed inner disc. This leads to a 'diagnostic diagram', as invented by Allen Shafter.[4]

One first measures the line centroid at each orbital phase, and derives the velocity shift by fitting with a sinusoid. One then repeats the line centroid calculations, progressively excluding more and more of the line centre (starting, for example, with the central ± 100 km s^{-1} and incrementing in steps of 100 km s^{-1}). Plotting (as a function of the amount excluded) the amplitude, phase and mean velocity of

---

*When looking at the limb of a star the line of sight is at a high angle, so it encounters much more opacity before attaining a given depth. Thus the photons come from higher up and are cooler, and so the limbs look fainter. Gravity darkening is a similar effect caused by the gravitational equipotentials being spaced differently in different parts of the Roche geometry, which affects the temperature structure of the star.

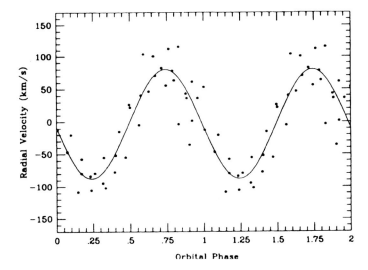

Fig. A.2: The radial velocity of the Hα line of AR And round its orbital cycle. The amplitude of the fitted sinusoid, K_1, is 83 ± 7 km s^{-1}. (Figure by Allen Shafter.[5])

the fitted sinusoid, together with an error estimate, enables one to understand the effect of the stream, and so deduce the true K_1 value. Ideally, once the disturbed line centre is excluded, the velocity amplitude and phase become constant, until in the extreme wings the signal becomes so weak that the error rises sharply. If the phasing of the binary is known, one can check that the fitted phase corresponds to that expected for white dwarf motion.

In practice, diagnostic diagrams don't always behave as desired. It seems that in many systems departures from symmetric line emission occur at all disc radii, perhaps due to a stream overflowing the initial impact with the disc and continuing towards the inner regions. Thus, using the disc lines yields K_1 in only a minority of systems, and is reliable only in eclipsing systems where one can easily check that the phase is correct. Since the mass function is very sensitive to such errors, depending on K_1 to the third power, the whole procedure is suspect. This hasn't prevented its use in dozens of research papers, including some of my own!

A.3 Measuring K_2

Measuring K_2 is, in principle, more straightforward, since absorption lines from the secondary can be observed in the red part of the spectrum. The sodium doublet (a pair of Na I lines at λ8183 and 8195) is commonly used.[6] However, the red-star features are often weak or obliterated in the glare of the disc, and the sodium doublet in particular can by contaminated by nearby lines.

192 Deriving the stellar masses   Appendix A

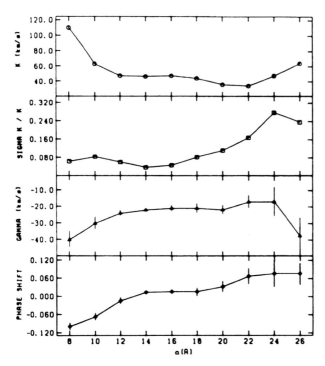

Fig. A.3: The 'diagnostic diagram' of SW UMa. When excluding medium amounts of line centre, K_1 (*upper panel*) is roughly constant near 47 km s^{-1}. Excluding larger amounts increases the error (*second panel*) whereas excluding smaller amounts leaves K_1 affected by line-centre disturbances. The absence of a phase shift (*bottom panel*) in the middle region confirms that 47 km s^{-1} is the best estimate of K_1. (Figure by Allen Shafter.[4])

Furthermore, such measurements will only give K_2 if the absorption line is uniformly distributed over the secondary star. Again, limb and gravity darkening can affect this. More seriously, irradiation from the primary can obliterate the line on the illuminated face, so that the observed absorption arises only from the rear. However, an equatorial belt of the secondary will be shadowed by the accretion disc, and so might still show absorption. In principle, corrections for all such factors can be calculated, although not yet with complete confidence. Doppler tomography of the absorption lines can help by showing the actual distribution of the line.

A similar method uses the *rotation* of the secondary rather than its orbital motion. If the secondary rotation is locked to the orbit (which tides will enforce) then the observed equatorial velocity, $v_{\rm eq}\sin i$, is given by

$$v_{\rm eq}\sin i = \frac{2\pi R_2}{P_{\rm orb}} \sin i$$

where R_2 is the volume-averaged radius of the secondary's Roche lobe. Since R_2 is a function of q and a only (see Box 2.4), this leads to a constraint on q and i that can complete the jigsaw puzzle. Again, though, deducing $v_{\rm eq}$ from an observed line profile requires a model of the line emission including the effects of limb darkening and irradiation.[6]

Appendix A   Deriving the stellar masses   193

A.4 Measuring M_1 directly

There are several methods of directly estimating the mass of the white dwarf. All are based on the fact that a white dwarf has a set relation between its mass and radius, which, for a cold, non-rotating helium white dwarf, is given by

$$R_{wd} = 0.779 \left(M^{-2/3} - M^{2/3} \right)^{1/2} 10^7 \text{ m}$$

where M is the white dwarf mass as a fraction of the Chandrasekhar mass of 1.44 $M_\odot$.[7]

Because of this, any accurate measure of R_{wd} will yield its mass. In an eclipsing system, the time of white-dwarf ingress or egress can give R_{wd} in the manner of Box 2.2, although other information is needed to first deduce the orbital speed of the secondary.

Given the mass–radius relation, the gravity on the surface of a white dwarf will be a function only of its mass. Since the gravity will determine the width of the white dwarf lines (see Section 3.1), modelling these can yield the mass. Similarly, the gravitational red-shift[†] of a white dwarf line can give the mass, the problem here being to separate the gravitational red-shift from the Doppler shift due to the motion of the binary with respect to Earth.

A last example of such methods applies to the magnetic cataclysmic variables. In these systems the accretion flow falls radially along field lines, not via a disc, and its energy is converted into heat suddenly when it hits the white dwarf. By measuring the temperature of the collision site one can deduce the gravitational energy liberated and hence the mass of the white dwarf, but again the results are dependent on the details of the shock region.[8]

The drawback of all these methods is that the white dwarfs in cataclysmic variables acquire a layer of accreted material, and it is not known how much this disturbs the star from the theoretical mass-radius relation.[9]

A.5 Estimating M_2 from P_{orb}

Given the problems with all methods of determining masses, it is often useful to fall back on an approximate estimate of M_2 deduced solely from the orbital period. The starting point is our previous expressions for the separation a and the secondary Roche-lobe radius R_2 (see Boxes 2.2 and 2.4)

$$a^3 = \frac{G(M_1 + M_2)P_{orb}^2}{4\pi^2} \quad \text{and} \quad R_2 = a\, 0.462 \left(\frac{q}{1+q} \right)^{1/3}$$

which can be manipulated to show that the mean density of the secondary is a function solely of P_{orb}

$$\bar{\rho} \equiv \frac{3M_2}{4\pi R_2^3} \approx 115\,000\, P_{orb}^{-2} \text{ kg m}^{-3}.$$

[†]The energy lost by a photon in climbing out of an intense gravitational field causes it to be red-shifted.

Now, if we know the structure of a red dwarf star — that is, how its radius depends on its mass — we can eliminate R_2 in the above equation and obtain M_2 as a function solely of $P_{\rm orb}$. This requires (1) a calibration of the mass–radius relation for red dwarfs, using single stars or wide binaries, and (2) the assumption that the same relation applies to the red dwarfs in cataclysmics. Both of these points have been debated at length. A widely used calibration for low-mass red dwarfs ($M < 1$ $M_\odot$) is

$$R = M^{0.867}$$

(both values in solar units) which leads to a secondary mass of

$$M_2 \approx 0.065 P_{\rm orb}^{5/4}\ M_\odot$$

where $P_{\rm orb}$ is in hours.[10] Concerning the second point, the current consensus is that the assumption is valid, and that the mass-radius relation of a red dwarf is not unduly distorted by its participation in a cataclysmic binary, though there are some anomalous systems that do appear to have distorted secondaries.[11]

For orbital periods greater than ≈ 9 hrs, the required density becomes too low to be satisfied by a dwarf star. Thus long-period cataclysmics must contain evolved secondaries that have expanded their outer layers in the manner of red giants. The above formulae then no longer apply.

Appendix B: Note on units and symbols

MASS-TRANSFER RATE: Given a solar mass of 2.0×10^{30} kg, 10^{-10} $M_\odot$ yr^{-1} = 6.3×10^{12} kg s^{-1} = 6.3×10^{15} g s^{-1}.

MAGNETIC FIELD STRENGTH: One Tesla equals 10^4 Gauss.

WAVELENGTH/ENERGY/TEMPERATURE: The common unit of wavelength, the Ångstrom (Å) is 10^{-10} m = 0.1 nm. For light, frequency $\nu = c/\lambda$. This can be converted into an energy using $E = h\nu$, and is often expressed in electron volts, eV, where 1 eV = 1.6×10^{-19} Joule. The cgs unit of energy, the erg, equals 10^{-7} J.

Thus a typical optical photon has λ = 5000 Å $\equiv$ 6.0×10^{14} Hz $\equiv$ 4.0×10^{-19} J $\equiv$ 2.5 eV. A 1-keV X-ray photon has a wavelength of 12.4 Å.

Energy can be converted to temperature using $E = kT$. Thus 100 000 K equates to 8.6 eV. Note, though, that this is not the peak of the blackbody distribution. A 100 000-K blackbody plotted against wavelength peaks at 290 Å (from Wien's law that $\lambda_{\text{peak}} T = 2.9 \times 10^{-3}$ m K), where a photon has an energy of 43 eV.

SYMBOLS USED IN THIS BOOK:

M_1	Primary star mass	M_2	Secondary star mass
a	Binary separation	q	Mass ratio $\equiv M_2/M_1$
i	Inclination	K	Orbital velocity
Σ	Surface density	c_s	Sound speed
ν	Viscosity	α	Viscosity parameter
L_1	Inner Lagrangian point	$\dot{M}$	Mass-transfer rate
P_{orb}	Orbital period	P_{spin}	Spin period
T	Temperature	r	Radius
M	Mass	L	Luminosity
E	Energy	J	Angular momentum
v	Velocity	H	Height
λ	Wavelength	F	Force

CONSTANTS USED IN THIS BOOK:

c	Speed of light	3.00×10^8 m s^{-1}
G	Gravitational constant	6.67×10^{-11} N m^2 kg^{-2}
k	Boltzmann's constant	1.38×10^{-23} J K^{-1}
h	Planck's constant	6.62×10^{-34} J s
e	Electron charge	1.60×10^{-19} C
m_e	Electron mass	9.11×10^{-31} kg
σ	Stefan–Boltzmann constant	5.67×10^{-8} J K^{-4} m^{-2} s^{-1}
m_p	Proton mass	1.67×10^{-27} kg
μ_0	Permeability of space	$4\pi \times 10^{-7}$ H m^{-1}
$M_\odot$	Solar mass	1.99×10^{30} kg
$R_\odot$	Solar radius	6.96×10^8 m
eV	Electron volt	1.60×10^{-19} J
pc	Parsec	3.09×10^{16} m

Appendix C: Time conventions

Civil time systems are based on Coordinated Universal Time (UTC, also called Greenwich Mean Time), in which one rotation of the Earth takes one day. Local civil time has a fixed offset from UTC, usually a whole number of hours ('summer time' or 'daylight saving time' may add or subtract another hour).

Astronomical observations, however, need to account for the time light takes to travel across the Solar System. This arises because a signal from a distant star will pass the Earth earlier or later depending on where the Earth is in its orbit around the Sun. Hence Earth times are often converted to 'heliocentric times', being the time when the signal would arrive at the Sun's centre. Given that light takes 8 minutes to travel the Sun–Earth distance, heliocentric corrections are needed for all observations that are accurate to minutes. In fact, the gravity of the planets (principally Jupiter) causes the Sun's location to change by 2–3 light seconds, so observations accurate to seconds should instead be corrected to 'barycentric time', being the time when the signal passes the barycentre of the Solar System.

Another complication is that Earth's rotation speed is not constant, so UTC, which is kept in step with Earth's rotation by the addition of 'leap seconds', is not a uniform timescale. Instead, where accuracy of a few seconds matters, one uses 'dynamical time' (previously called 'ephemeris time') in which a day equals 86400 secs as measured by atomic clocks. Currently, Terrestrial Dynamical Time (abbreviated TDT) differs from UTC by ~ 1 min, owing to accumulated leap seconds. Barycentric Dynamical Time (TDB) is dynamical time referred to the Solar System barycentre.

The civil calender, involving months of different duration, leap years, calender reforms, and so on, is also awkward for astronomical purposes. Instead, one often uses 'Julian Day Number', which is a simple count of the days since noon Greenwich Mean Time on 1 January 4713 BC. 'Julian Date' is the Julian Day Number followed by the fraction of the day elapsed since the preceding noon. For example, 0 UT on 1 January 2000 was Julian Date 2451544.5000. Since Julian Date involves large numbers, the leading digits are often omitted, for example when labelling figures.

Julian Date can be used with any of the above time systems so, for example, 'HJD' means a heliocentric time expressed as a Julian Date, whereas 'TDB(JD)' means a barycentric dynamical time expressed as a Julian Date.

The formulae for interconversion of different timescales are too involved to be presented here. However, the *Astronomical Almanac*, published every year by the US and UK governments, is the definitive account, while *Astronomy with your Personal Computer*, by Peter Duffett-Smith (Cambridge University Press, 1990) describes algorithms for making the calculations.

Appendix D: Variable star nomenclature

Variable stars are named according to the following conventions, redolent of the topic's long history. Very bright variables might possess traditional names (for example Algol) or have been included amongst the brightest stars in each constellation labelled with Greek letters, Roman letters, or Flamsteed numbers (for example δ Cephei). If so, these names are used.

Where there was no existing name, the first variable discovered in each constellation was labelled 'R' along with the genitive form of the constellation (e.g. R Coronae Borealis), most often abbreviated (e.g. R CrB). Subsequent variables were labelled with S, T ... Z. After these nine, further variables were given a two-letter prefix, with the rule that the second letter be no earlier in the alphabet than the first. This two-letter system also began with R, so the next stars were labelled RR, RS, RT ... RZ, followed by SS, ST ... SZ. After ZZ comes AA, AB, AC etc. However the letter J is never used. After 334 stars one arrives at QZ and the end of the system. Further variables, more prosaically, attract a number starting at 335, preceded by V for variable (e.g. V335 Cyg, V336 Cyg, etc).

In a similar way, the first stars observed by X-ray satellites were labelled with an X (e.g. Cyg X-1, Cyg X-2, etc). A separate X-ray name was needed since the satellites had poorer resolution than optical telescopes, and it was often unclear which optical star was the counterpart of the X-ray source. Nowadays, X-ray sources are assigned a catalogue number based on an abbreviation of the satellite name and the star's coordinates (for example a star first seen with the *EXOSAT* satellite at right ascension $07^h 48^{min}$ and declination $-67.6°$ was designated EXO 0748-676). Once an optical counterpart has been identified, a variable-star name will also be assigned, though X-ray astronomers may prefer to use the X-ray name. Variable-star names are not assigned to variables found in globular clusters or other galaxies.

Appendix E: Variable star organisations

The **American Association of Variable Star Observers** (AAVSO) is the largest variable-star organisation. Based in the US, but with members in 40 countries, it acts as the international clearing house for amateur observations of variable stars. It provides star charts, lightcurves, e-mail alert notices, circulars, and material for the novice observer. Particularly useful is a web-based facility that plots lightcurves of any star in the AAVSO database. Web site: `http://www.aavso.org/` e-mail: `aavso@aavso.org` Address: 25 Birch Street, Cambridge, MA 02138-1205, USA.

The **Variable Star Section** of the **Royal Astronomical Society of New Zealand** (VSS RASNZ) is the foremost variable-star organisation in the southern hemisphere, publishing charts and circulars for southern variables. Web site: `http://www.rasnz.org.nz/` e-mail: `varstar@voyager.co.nz` Address: PO Box 3093, Greerton, Tauranga, New Zealand.

198   List of cataclysmic variables   Appendix F

The **Center for Backyard Astrophysics** (CBA) is a network of observers equipped with CCD cameras. By concentrating on a few targets, and by combining observations from different places to fill in gaps caused by cloud or daylight, it aims for dense and continuous coverage of variable-star lightcurves. The result is a steady stream of papers in the professional literature. Web site: http://cba.phys.columbia.edu e-mail: info@cba.phys.columbia.edu

The **Variable Star Network** (VSNet) runs e-mail lists for exchanging information on the status of variable stars and observing campaigns. It also maintains a database of observations, including those contributed by the Variable Star Observers League of Japan. Web site: http://www.kusastro.kyoto-u.ac.jp/vsnet/

The **Variable Star Section** of the **British Astronomical Association**.
Web site: http://www.ast.cam.ac.uk/~baa/ Address: BAA Assistant Secretary, Burlington House, Piccadilly, London, W1V 9AG, UK.

The **Association Francaise des Observateurs D'Etoiles Variables**. Web site: http://cdsweb.u-strasbg.fr/afoev/ e-mail: afoev@astro.u-strasbg.fr Address: AFOEV, c/o E. Schweitzer, 16 rue de Plobsheim, 67100 Strasbourg, France.

Appendix F: List of cataclysmic variables

The following is a list of the objects discussed in this book. For a fuller list of 318 cataclysmic variables with known or suspected orbital periods, complete with literature references, see the catalogue by Hans Ritter & Uli Kolb published in *Astronomy & Astrophysics Supplement*, 1998, vol 129, p83, and available at http://www.mpa-garching.mpg.de/Binary/ukolb/ed6.html. A list of all 1020 known cataclysmic variables (as of 1996) is presented by Downes, Webbink & Shara in the *Publications of the Astronomical Society of the Pacific*, 1997, vol 109, p345.

The data presented below are from Ritter & Kolb, supplemented by the AAVSO database and Warner (1995). Abbreviations used are: NL, novalike; VY, VY Scl low states; SW, SW Sex star; ZC, Z Cam star; DN, dwarf nova; SU, SU UMa star; ER, ER UMa star; WZ, WZ Sge star; AC, AM CVn star; AM, AM Her star; IP, intermediate polar; SH, permanent superhumps; IH, infrahumps; Na, fast nova; Nb, slow nova; Nr, recurrent nova; Ecl, eclipsing system. The magnitudes quoted are typical values (quiescence of DN, standstill of ZC and high state of VY), but are indicative only because they vary. The figures in brackets are: DN, typical outburst magnitude; ZC, outburst range; VY, typical low-state magnitude; AM, field strength; IP, spin period. For novae the bracket gives the year of outburst and peak magnitude, with the descriptors 'osc' for nova oscillations and 'dm' for a dust minimum, where seen. The distances quoted are indicative only because they are uncertain. For lightcurves of these objects use the AAVSO's 'lightcurve generator' at http://www.aavso.org/.

Appendix F   List of cataclysmic variables   199

Name	Coordinates	P_{orb}(hrs)	mag	dist (pc)	Notes
RS Oph	17^h50^m $-06°42'$	460 d	11.5	1800	Nr (5); red-giant companion
GK Per	03^h31^m $+43°54'$	47.92	13.2	340	DN (10) IP (351 s) Na (1901, 0.2, osc)
V723 Cas	01^h05^m $+54°00'$	16.64	18		Nb (1995, 6.7)
AE Aqr	20^h40^m $-00°52'$	9.88	11	140	NL; IP (33 s); likely propeller
RU Peg	22^h14^m $+12°42'$	8.99	13	174	DN (9.0)
Z Cam	08^h25^m $+73°06'$	6.96	11.4	173	DN ZC (10.0–13.0)
SS Cyg	21^h42^m $+43°35'$	6.60	11.9	166	DN (8.2); brightest dwarf nova
AH Her	16^h44^m $+25°15'$	6.19	12.5	250	DN ZC (11.0–14.5)
XY Ari	02^h56^m $+19°26'$	6.06		200	IP (206 s) DN; Ecl; behind H$_2$ cloud
TX Col	05^h43^m $-41°01'$	5.72	15.7	550	IP (1911 s)
RW Tri	02^h25^m $+28°05'$	5.56	12.7	224	NL; Ecl
TV Col	05^h29^m $-32°49'$	5.49	13.8		IP (1910 s); IH; 'flares'; Ecl
V705 Cas	23^h41^m $+57°31'$	5.47	18		Na (1993, 6.5, dust minimum)
PQ Gem	07^h51^m $+14°44'$	5.18	14.3		IP (833 s, ≈ 15 MG)
HR Del	20^h42^m $+19°09'$	5.14	12		Nb (1967; 3.5)
EX Dra	18^h04^m $+67°54'$	5.04	15		DN (13); Ecl
RX And	01^h04^m $+41°17'$	5.04	11.8		DN ZC (10.5–14.0)
FO Aqr	22^h17^m $-08°21'$	4.85	13.5	325	IP (1254 s); grazing eclipse
UX UMa	13^h36^m $+51°54'$	4.72	12.7		NL; Ecl
IX Vel	08^h15^m $-49°13'$	4.65	9.8		NL; the brightest novalike
DQ Her	18^h07^m $+45°51'$	4.65	14.2		IP (71 s); Ecl; Na (1934, 1.4, dm)
SS Aur	06^h13^m $+47°44'$	4.39	15	200	DN (10.5)
U Gem	07^h55^m $+22°00'$	4.25	14	96	DN (9.0); grazing eclipse
RX1313–32	13^h13^m $-32°59'$	4.19	16	200	AM (56 MG)
LS Peg	21^h51^m $+14°06'$	4.19	13.0		NL SW
V405 Aur	05^h58^m $+53°53'$	4.15	14.6		IP (545 s)
AR And	01^h45^m $+37°56'$	3.91	16.0		DN (11.5)
KR Aur	06^h15^m $+28°35'$	3.91	13.2		NL VY (18)
IP Peg	23^h23^m $+18°24'$	3.80	15.6		DN (12.2); Ecl
QQ Vul	20^h05^m $+22°39'$	3.71	15.6		AM (34 MG)
AO Psc	22^h55^m $-03°10'$	3.59	13.6	420	IP (805 s)
PX And	00^h30^m $+26°17'$	3.51	15.3		NL SW; Ecl; SH IH
V2400 Oph	17^h12^m $-24°14'$	3.40	14.3		IP (927 s, ≈ 15 MG); discless
V1223 Sgr	18^h55^m $-31°09'$	3.37	13.3	600	IP (746 s); VY (17)
V1432 Aql	19^h40^m $-10°25'$	3.37	15	230	asynch AM
BY Cam	05^h42^m $+60°51'$	3.35	14.8	190	asynch AM
V1315 Aql	19^h13^m $+12°18'$	3.35	14.5		NL SW; Ecl

Name	Coordinates	P_{orb}(hrs)	mag	dist (pc)	Notes
V1500 Cyg	21^h11^m $+48°09'$	3.35	20	1200	asynch AM; Na (1975, 2.2)
V603 Aql	18^h48^m $+00°35'$	3.31	11.5		Na (1918, −1.1, osc); SH IH
TT Ari	02^h06^m $+15°17'$	3.30	10.6		NL VY (15.6); SH IH
DW UMa	10^h33^m $+58°46'$	3.28	14.5		NL SW VY (16); IH; Ecl
SW Sex	10^h15^m $-03°08'$	3.24	15		NL SW; Ecl
MV Lyr	19^h07^m $+44°01'$	3.19	12.2		NL VY (16); SH
NN Ser	15^h52^m $+12°54'$	3.12	16.6		Detached binary
AM Her	18^h16^m $+49°52'$	3.09	13.0	75	AM (13 MG); VY (15.5)
QS Tel	19^h38^m $-46°12'$	2.33	15.5	300	AM (47 MG); VY (17.4)
HU Aqr	21^h07^m $-05°17'$	2.08	15.3	190	AM (35 MG); VY (20); Ecl
DV UMa	09^h46^m $+44°46'$	2.06	18.5		DN SU (15); Ecl
AR UMa	11^h15^m $+42°58'$	1.93	13.8		AM (230 MG); VY
AN UMa	11^h04^m $+45°03'$	1.91	14.5		AM (29 MG); VY (19)
ST LMi	11^h05^m $+25°06'$	1.90	15.5	128	AM (12 MG)
MR Ser	15^h52^m $+18°56'$	1.89	15		AM (28 MG)
CD Ind	21^h15^m $-58°40'$	1.85	16.5		asynch AM
SU UMa	08^h12^m $+62°36'$	1.83	14.4		DN SU (12)
WX Hyi	02^h09^m $-63°18'$	1.80	14.4		DN SU (11.5)
Z Cha	08^h07^m $-76°32'$	1.79	15.8	130	DN SU (12); Ecl
VW Hyi	04^h09^m $-71°17'$	1.78	13.8	65	DN SU (9)
HT Cas	01^h10^m $+60°04'$	1.77	16.5		DN SU (11); Ecl
T Pyx	09^h04^m $-32°22'$	1.75	15.2	1000	Nr (6.5)
V834 Cen	14^h09^m $-45°17'$	1.69	14.5	86	AM (23 MG)
VV Pup	08^h15^m $-19°03'$	1.67	15	145	AM (31 MG)
EX Hya	12^h52^m $-29°14'$	1.64	13	105	IP (4021 s); DN (9.5); Ecl
ER UMa	09^h47^m $+51°54'$	1.53	15		DN SU ER (12.5)
OY Car	10^h06^m $-70°14'$	1.51	15.5		DN SU (12); Ecl
V1494 Aql	19^h23^m $+04°57'$	1.50		3600	Na (1999, 3.5, oscillations)
V1159 Ori	05^h28^m $-03°33'$	1.49	15		DN SU ER (13)
RZ LMi	09^h51^m $+34°07'$	1.40	16.8		DN SU ER (14.4)
EG Cnc	08^h43^m $+27°51'$	1.38	18.5		DN SU WZ (12); 'echos'
SW UMa	08^h36^m $+53°28'$	1.36	16.7		DN SU (10)
WZ Sge	20^h07^m $+17°42'$	1.36	15	48	DN SU WZ (7.8); IP? (28 s)
EF Eri	03^h14^m $-22°35'$	1.35	15	94	AM (8 MG); VY (17)
V803 Cen	13^h23^m $-41°44'$	0.45	14.7		AC DN SU (13.2)
AM CVn	12^h34^m $+37°37'$	0.29	14.1		AC NL; SH IH

Bibliography

The essential reference work on cataclysmic variables is Warner's *Cataclysmic Variable Stars* (Cambridge University Press, 1995). It contains a longer and deeper-level account than this book and is packed with a wealth of observational detail on individual systems. In the bibliography below, it goes without saying that 'Warner' is the prime resource for further reading.

The graduate-level text *Accretion Power in Astrophysics* by Frank, King and Raine (Cambridge University Press, 1992) is an excellent introduction to the theory of accretion, with much that is relevant to cataclysmic variables.

Accretion Discs in Compact Stellar Systems, edited by Wheeler (1993, World Scientific Press) is highly recommended as an account of all aspects of accretion discs, and is also at a higher level than this book. It is abbreviated to ADCSS below.

Every year or two a newly published conference proceedings provides a snapshot of the state of current research. The latest of these, particularly notable as a collection of review articles, arose from the 1999 Oxford conference in honour of Brian Warner's 60th birthday, and is published in *New Astronomy Reviews*, 2000, vol 44, pp 1–178, edited by Charles, King and O'Donoghue. Similarly, the forthcoming *Astro-tomography*, edited by Boffin, Steeghs and Cuypers (Springer–Verlag), is recommended for reviews of tomographic techniques applied to cataclysmic variables.

NASA's *Astrophysics Data System* lists essentially all astronomy research papers! Accessed through the web site http://ads.harvard.edu/, its abstract service is *the* way of keeping abreast of developments reported in journals or at conferences, as well as the easiest way of searching for and accessing older material.

The following list of references is not exhaustive, but records the sources of figures and is a starting point for further reading. The abbreviations are: MNRAS, *Monthly Notices of the Royal Astronomical Society*; ApJ, *Astrophysical Journal*; AJ, *Astronomical Journal*; A&A, *Astronomy and Astrophysics*; PASP, *Publications of the Astronomical Society of the Pacific*; PASJ, *Publications of the Astronomical Society of Japan*; ASP Conf, *Astronomical Society of the Pacific conference series*; and CUP, *Cambridge University Press*.

Chapter 1:
1. 'Warner' Chapter 1 gives an account of the history of cataclysmic variable research

Chapter 2:
1. James Kaler's *Stars*, (Scientific American Library, No 39, 1998) and the more mathematical *An Introduction to the Theory of Stellar Structure and Evolution*, by D. Prialnik (CUP, 2000) are good introductions to stars
2. Unpublished data, but see Catalán M. S. et al. MNRAS, 269, 879, for a study of NN Ser
3. See, for example, *Newtonian Mechanics*, A. P. French (W. W. Norton, 1971)
4. Allan A., Hellier C., Ramseyer T. F., 1996, MNRAS, 282, 699
5. Lubow S. H., Shu F. H., 1975, ApJ, 198, 383, is the classic account
6. Patterson J. et al. PASP, December 2000
7. Harrop-Allin M. K., Warner B., 1996, MNRAS, 279, 219
8. Wood J. et al. 1986, MNRAS, 219, 629, is a classic analysis of an eclipse
9. Patterson J., 1980, ApJ, 241, 235

Chapter 3:
1. D. Emerson's *Interpreting Astronomical Spectra* (Wiley, 1996) is a detailed and mathematical account of stellar spectra.
2. Hellier C., Buckley D. A. H., 1993, MNRAS, 265, 766
3. Thomas H.-C., Beuermann K., Burwitz V. et al. 2000, A&A, 353, 646
4. Hessman F. V., Robinson E. L., Nather R. E., Zhang E.-H., 1984, ApJ, 286, 747
5. See the review by G. Shaviv and R. Wehrse in ACDSS
6. See Frank J, King A. R., Raine D., *Accretion Power in Astrophysics* (CUP, 1992), chapter 5 for a fuller account of disc temperatures
7. See the chapter by K. Horne in ADCSS for a review of eclipse mapping
8. Marsh T. R. et al. 1990, ApJ, 364, 637
9. Spruit H. C., Rutten R. G. M., 1998, MNRAS, 299, 768
10. Horne K., Marsh T. R., 1986, MNRAS, 218, 761
11. The original paper was Marsh T. R., Horne K., 1988, MNRAS, 235, 269. See also Kaitchuck R. H. et al. 1994, ApJ Supp, 93, 51, and the chapter by Robinson E. L., Marsh T. R., Smak J. I., in ADCSS

Chapter 4:
1. Ritter H., Kolb U., 1998, A&A Suppl, 129, 83
2. For a fuller review of evolution see King A. R., 1988, *Quarterly Journal of the Royal Astronomical Society*, 29, 1, and updates including King A. R., 1998, ASP Conf, 137, 174, and King A. R., 2000, New Astronomy Reviews, 44, 167. Another good reference is Livio M., in Saas-Fee Advanced Course 22 (Springer–Verlag, 1994)
3. Harrison T. E. et al., 1999, ApJ Lett, 515, L93

Chapter 5:
1. See Warner B., *High Speed Astronomical Photometry* (CUP, 1988)
2. Osaki Y., 1974, PASJ, 26, 429
3. Rutten R. G. M., Kuulkers E., Vogt N., van Paradijs J., 1992, A&A, 265, 159
4. Smak J., 1984, Acta Astronomica, 34, 93
5. Lynden-Bell D., Pringle J. E., 1974, MNRAS, 168, 603
6. Shakura N. I., Sunyaev R. A., 1973, A&A, 24, 337
7. See, e.g., the review Hawley J. F., Balbus S. A., 1998, ASP Conf, 137, 273

8. See, e.g., the review by Cannizzo J. K., in ADCSS for a fuller account of the thermal instability mechanism and the topic's history; for a lower-level treatment see Cannizzo J. K., Kaitchuck R. H., 1992, Scientific American, 266, 42
9. Mineshige S., Osaki Y., 1985, PASJ, 37, 1
10. Webb N. A. et al., 1999, MNRAS, 310, 407
11. Baptista R., Catalán M. S., in *Cataclysmic Variables: a 60th Birthday Symposium in Honour of Brian Warner* (Elsevier, 2000)
12. Horne K., la Dous C. A., Shafter A. W., in *Accretion-Powered Compact Binaries* (CUP, 1990)
13. Ichikawa S., Osaki Y., 1992, PASJ, 44, 15
14. Ritter H., Kolb U., 1998, A&A Suppl, 129, 83

Chapter 6:
1. Vogt N., 1982, ApJ, 252, 653
2. Patterson J. et al., 2000, PASP, Dec issue
3. Whitehurst R., 1988, MNRAS, 232, 35
4. Hirose M., Osaki Y., 1990, PASJ, 42, 135
5. See Whitehurst R., King A., 1991, MNRAS, 249, 25, and Lubow S. H., 1992, ApJ, 401, 317
6. Ritter H., Kolb U., 1998, A&A Suppl, 129 83
7. See Patterson J., 1998, PASP, 110, 1132, and also Murray J., 2000, MNRAS, 314, L1
8. O'Donoghue D., 1990, MNRAS, 246, 29
9. Patterson J. et al., 1993, PASP, 105, 69
10. Patterson J. et al., 1995, PASP, 107, 1183
11. Patterson J. et al., 2000, PASP, Dec issue
12. Rolfe D., Haswell C. A., Patterson J., 2000, MNRAS, in press
13. See Osaki Y., 1996, PASP, 108, 39 for a review of this topic
14. Ichikawa S., Hirose M., Osaki Y., 1993, PASJ, 45, 243
15. Hameury J.-M., Lasota J.-P., Warner B., 2000, A&A, 353, 244
16. Robertson J. W., Honeycutt R. K., Turner G. W., 1995, PASP, 107, 443
17. Patterson J. et al., 1998, PASP, 110, 1290
18. Patterson J., 1999, in *Disk Instabilities in Close Binary Systems*, Frontiers Science Series No. 26 (Universal Academy Press)
19. Hellier C., 1993, MNRAS, 264, 132
20. Steeghs D., Stehle R., 1999, MNRAS, 307, 99
21. Steeghs D., Harlaftis E. T., Horne K., 1997, MNRAS, 290, 28; see also Steeghs D., Harlaftis E., 1997, Sky & Telescope, 94, 20
22. Harlaftis E. T., Steeghs D., Horne K. et al., 1999, MNRAS, 306, 348

Chapter 7:
1. See, e.g., Drew J. E., Kley W. in ADCSS for a fuller account
2. See, e.g., Meyer F., Liu B. F., Meyer-Hofmeister E., 2000, A&A, 361, 175, and Meyer F., 1999 in *Disk Instabilities in Close Binary Systems*, Frontiers Science Series No. 26 (Universal Academy Press)
3. See Drew J. E., Proga D., 2000, New Astronomy Reviews, 44, 21 for a review
4. Wheatley P. J., Mauche C. W., Mattei J., work in preparation
5. Hassall B. J. M., Pringle J. E., Verbunt F., 1985, MNRAS, 216, 353
6. Mason K. O., Córdova F. A., Watson M. G., King A. R., 1988, MNRAS, 232, 779

7. Hellier C., Mason K.O., Cropper M., 1990, MNRAS, 242, 250
8. Parmar A.N., White N.E., Giommi P., Gottwald M., 1986, ApJ, 308, 199
9. Armitage P.J., Livio M., 1998, ApJ, 493, 898
10. Thorstensen J.R. et al., 1991, AJ, 102, 272
11. See Hellier C., 2000, New Astronomy Reviews, 44, 131, and Horne K., 1999, ASP Conf, 157, 349, for contrary opinions
12. Knigge C. et al., 2000, ApJ Lett, 539, L49
13. See Hellier C., Robinson E.L., 1994, ApJ Lett, 431, L107, and Hellier C., 1998, PASP, 110, 420
14. Patterson J., 1999, in *Disk Instabilities in Close Binary Systems*, Frontiers Science Series No. 26 (Universal Academy Press)
15. Hellier C., 1996, ApJ, 471, 949
16. Taylor C.J., Thorstensen J.R., Patterson J., 1999, PASP, 111, 184

Chapter 8:
1. Schmidt G.D., 1999, ASP Conf, 157, 207
2. Cropper M., 1990, Space Science Reviews, 54, 195 is the classic review of AM Her stars
3. Kube J., Gänsicke B.T., Beuermann K., 2000, A&A, 356, 490
4. Harrop-Allin M.K. et al., 1999, MNRAS, 308, 807
5. Schwope A.D., Mantel K.-H., Horne K.D., 1997, A&A, 319, 894, and Heerlein C., Horne K., Schwope A.D., 1999, MNRAS, 304, 145
6. E.g. Mason K.O., 1985, Space Science Reviews, 40, 99
7. Thomas H.-C., Beuermann K., Burwitz V. et al., 2000, A&A, 353, 646
8. Tapia S., 1977, ApJ Lett, 212, L125
9. Potter S., 2000, MNRAS, 314, 672, and Cropper M.S., 1986, MNRAS, 222, 225
10. Schwope A.D., Beuermann K., Jordan S., Thomas H.-C., 1993, A&A, 278, 487
11. Beuermann K., in *High Energy Astronomy & Astrophysics*, Tata Institute of Fundamental Research, Mumbai, India, (Sangam Books, 1998)
12. Heise J. et al., 1985, A&A, 148, L14
13. Patterson J., Skillman D.R., Thorstensen J., Hellier C., 1995, PASP, 107, 307
14. Ramsay G., Potter S., Cropper M. et al., 2000, MNRAS, 316, 225
15. Schmidt G.D., Stockman H.S., 1991, ApJ, 371, 749

Chapter 9:
1. Patterson J., 1994, PASP, 106, 209, is a major review of DQ Her stars
2. Buckley D.A.H. et al., 1997, MNRAS, 287, 117
3. E.g. Wynn G.A., King A.R., 1995, MNRAS, 275, 9
4. King A.R., Wynn G.A., 1999, MNRAS, 310, 203
5. See Hellier C., 1999, ASP Conf, 157, 1, for more details
6. See Wynn G.A., 2000, New Astronomy Reviews, 44, 75, for a review
7. Hellier C., 1993, MNRAS, 265, L35
8. See Hellier C., 1991, MNRAS, 251, 693, and Wynn G.A., King A.R., 1992, MNRAS, 255, 83
9. Hellier C., 1998, Advances in Space Research, 22, 973
10. Murray J.R. et al., 1999, MNRAS, 302, 189
11. Patterson J. et al., 1998, PASP, 110, 415
12. See Li J., 1999, ASP Conf, 157, 235, and references therein
13. Hellier C., 1997, MNRAS, 291, 71

14. Hellier C., Mukai K., Beardmore A.P., 1997, MNRAS, 292, 397
15. Hellier C. et al., 2000, MNRAS, 313, 703
16. Rosen S.R., Mason K.O., Córdova F.A., 1988, MNRAS, 231, 549
17. Hellier C., Cropper M., Mason K.O., 1991, MNRAS, 248, 233
18. Hellier C., Mukai K., Ishida M., Fujimoto R., 1996, MNRAS, 280, 877
19. Siegel N., Reinsch K., Beuermann K. et al., 1989, A&A, 225, 97
20. Allan A. et al., 1996, MNRAS, 279, 1345
21. Warner B., Cropper M., 1984, MNRAS, 206, 261, but note that this paper preceded the discovery of the true spin period
22. See Zhang E. et al., 1995, ApJ, 454, 447, and references therein
23. Warner B., 1986, MNRAS, 219, 347
24. Nather R.E., 1978, PASP, 90, 477
25. Wynn G.A., King A., Horne K., 1997, MNRAS, 286, 436; see also Welsh W.F., 1999, ASP Conf, 157, 357 for a review
26. Patterson J., 1979, ApJ, 234, 978
27. Wynn G.A., Leach R., King A.R., 2000, MNRAS, submitted
28. See Kolb U., 1995, ASP Conf, 85, 440, and Wickramasinghe D.T., Li J., Wu K., ASP Conf, 85, 452, for reviews
29. See Beardmore A.P. et al., 2000, MNRAS, 315, 307, and Cropper M. et al., 1999, MNRAS, 306, 684, together with references therein

Chapter 10:
1. Warner B., 1976, IAU Symposium 73, 85, and Patterson J., 1981, ApJ Supp, 45, 517
2. Horne K., Stiening R.F., 1985, MNRAS, 216, 933
3. Robinson E.L., Nather R.E., 1979, ApJ Supp, 39, 461
4. Patterson J., 1981, ApJ Supp, 45, 517, is a good review of DNOs and QPOs
5. Warner B., O'Donoghue D., Wargau W., 1989, MNRAS, 238, 73
6. Warner B., 1995, ASP Conf 85, 343
7. Córdova F.A., Chester T.J., Mason K.O. et al., 1984, ApJ, 278, 739
8. E.g., Patterson J., 1991, PASP, 103, 1149
9. Watson M.G., King A.R., Osborne J., 1985, MNRAS, 212, 917
10. Hellier C., Livio M., 1984, ApJ Lett, 424, L57
11. Larsson S., 1992, A&A, 265, 133
12. Chanmugam G., 1995, ASP Conf, 85, 317
13. Larsson S., 1995, ASP Conf, 85, 311
14. Warner B., van Zyl L., 1998, IAU Symposium 185, 321
15. van Zyl L. et al., 1999, Baltic Astronomy, 9, 231

Chapter 11:
1. See Starrfield S., Truran J.W., Sparks W.M., 2000, New Astronomy Reviews, 44, 81, and references therein
2. Iijima T., Rosino L., della Valle M., 1998, A&A, 338, 1006
3. Livio M., in Saas-Fee Advanced Course 22 (Springer–Verlag, 1994), addresses many of the issues from this section
4. For a review see Lamb D.Q., 2000, ApJ Supp, 127, 395
5. See Gänsicke B.T., van Teeseling A., Beuermann K., Reinsch K., New Astronomy Reviews, 44, 143, for a review
6. Hjellming M.S., Webbink R.F., 1987, ApJ, 318, 794

Chapter 12:
1. Rosenzweig P., Mattei J. A., Kafka S. et al., 2000, PASP, 112, 632
2. King A. R., Cannizzo J. K., 1998, ApJ, 499, 348, and Leach R., Hessman F. V., King A. R., Stehle R., Mattei J., 1999, MNRAS, 305, 225
3. Hessman F. V., 2000, New Astronomy Reviews, 44, 155
4. Webb N., Naylor T., Jeffries R. D., 2000, in preparation
5. Warner B., 1988, Nature, 336, 129
6. Baptista R., Catalán M. S., Costa L., 2000, MNRAS, 316, 529; the EX Hya data are from Hellier C., Sproats L., 1992, Information Bulletin on Variable Stars, 3724
7. Bianchini A., 1990, AJ, 99, 1941
8. Applegate J. H., 1992, ApJ, 385, 621
9. Hellier C., Buckley D. A. H., 1993, MNRAS, 265, 766
10. Warren J. et al., 1993, ApJ Lett, 414, L69
11. Shara M. M., Livio M., Moffat A. F. J., Orio M., 1986, ApJ, 311, 163, and Livio M., Shara M. M., 1987, ApJ, 319, 819
12. Warner B., 1995, *Cataclysmic Variable Stars* (CUP, 1995)
13. Wu K., Wickramasinghe D. T., Warner B., 1995, Pub. Astr. Soc. Australia, 12, 60

Chapter 13:
1. Djurašević G., Zakirov M., Erkapić S., 1999, A&A, 343, 894
2. Richards M. T., Mochnacki S. W., Bolton C. T., 1988, AJ, 96, 326
3. See Charles P. A., Seward F. D., *Exploring the X-ray Universe* (1995, CUP) for an introductory account of X-ray binaries; the volume *X-ray Binaries* edited by Lewin W. H. G., van Paradijs J., van den Heuvel E. P. (CUP, 1997) is the definitive guide
4. King A. R., Kolb U., Szuszkiewicz E., 1997, ApJ, 488, 89
5. Muno M. P., Morgan E. H., Remillard R. A., 1999, ApJ, 527, 321
6. See Bradley Peterson's *An Introduction to Active Galactic Nuclei* (CUP, 1997) or Julian Krolik's *Active Galactic Nuclei* (Princeton University Press, 1998)
7. Bell K. R., Lin D. N. C., Hartmann L. W., Kenyon S. J., 1995, ApJ, 444, 376
8. Lee Hartmann's *Accretion Processes in Star Formation* (CUP, 1998) is an excellent treatment of star formation and of accretion in general

Appendix A:
1. See Horne K., 1985. MNRAS, 213, 129, and Wood J. et al. 1986, MNRAS, 219, 629
2. See, e.g., Orosz J. A. et al., 1998, ApJ, 499, 375 for a 'case study'
3. Sion E. M. et al., 1998, ApJ, 496, 449; see also Sion E. M., 1999, PASP, 111, 532 for a review of white dwarfs in cataclysmic variables
4. Shafter A. W., Szkody P., Thorstensen J. R., 1986, ApJ, 308, 765
5. Shafter A. W., Veal J. M., Robinson E. L., 1995, ApJ, 440, 853
6. E.g., Catalán M. S., Schwope A. D., Smith R. C., 1999, MNRAS, 310, 123
7. Nauenberg M., 1972, ApJ, 175, 417
8. E.g., Cropper M., Ramsay G., Wu K., 1998, MNRAS, 293, 222
9. See, e.g., Koester D., Schönberner D., 1986, A&A, 154, 125
10. From Warner B., *Cataclysmic Variable Stars* (1995, CUP)
11. See Beuermann K., 2000, New Astronomy Reviews, 44, 93, and Kolb U., Baraffe I., 2000, New Astronomy Reviews, 44, 99, for recent discussions of the red star

Object Index

Accounts including a figure have italic page numbers

Algol		3, *182*
AR	And	*193*
PX	And	*107*
RX	And	*151*
V603	Aql	161
V1315	Aql	*108*
V1432	Aql	123, *124*
V1494	Aql	165, *166*
AE	Aqr	*147, 148*
FO	Aqr	104, *134*, 136, *142*, *146*
HU	Aqr	112, *113*, 115
TT	Ari	*172*
XY	Ari	21, *23*, *137–140*
KR	Aur	*152*
SS	Aur	50
V405	Aur	*144*
BY	Cam	124
Z	Cam	*74*, 82, *83*
OY	Car	55, *56*, 57
HT	Cas	*152*
V705	Cas	165
V723	Cas	*164*
V834	Cen	*158*, 159
V1025	Cen	129
Z	Cha	*28, 29, 40*
EG	Cnc	90
TV	Col	*34*, 93, 176, *177*
TX	Col	135
SS	Cyg	2, 5, 7, *37*, 55, *69–71*, *101*, 155, *156*, *174*
V1057	Cyg	188
V1500	Cyg	124, 125, 161, *162*
HR	Del	161, *162*
EX	Dra	68, *175*
EF	Eri	*116*, *122*, 149, *159*
PQ	Gem	*9, 11*, 149
U	Gem	29, *30*, *55–59*, 87, *152*, *176*, 192
AH	Her	155
AM	Her	*122, 123*, 149, *173*
DQ	Her	145, *147*
EX	Hya	129, *140*, *143*, 144, *175*
VW	Hyi	6, 75, *76*
WX	Hyi	*103*
CD	Ind	*124*
GW	Lib	159, *160*
RZ	LMi	89
ST	LMi	113, *116*, 119
RS	Oph	167
V2400	Oph	127, *128*, *129*, 135, 149
V1159	Ori	*83, 84*
IP	Peg	67, *94*, 95
LS	Peg	108
RU	Peg	*154*
GK	Per	*157*, *164*, *176*
TY	PsA	*154*
AO	Psc	*141*
VV	Pup	113, *116*, *159*
T	Pyx	167
MR	Ser	*120*
NN	Ser	15, *16*, 17, 19, 21
WZ	Sge	29, *30*, *42*, *44*, 53, 99, 148
V1223	Sgr	*145*
RZ	Tau	*182*
QS	Tel	*177*
AN	UMa	*159*
AR	UMa	110, 111
DV	UMa	*76*
DW	UMa	105
IY	UMa	27, *84*, 85
SU	UMa	*172*
SW	UMa	*194*
QQ	Vul	116
EXO 0748–676		*104*, 168, *169*
RX 1313–32		*35*, *118*
X1624–69		*153*
X1636–53		*169*
GRS 1915+105		185, *186*
GS 2000+25		*183*

Index

accretion discs
   flare angle   105–106, 145, 179
   precesion   75–79, 92
   radii   28, 57–59, 85, 87
   spectra   34–37, 69
   temperatures   34, 39–40
Algol binaries   22, 181–182
AM CVn stars   54
AM Her stars   109–126, 136, 137, 158–159, 172–173, 177

Balbus–Hawley instability   61–63
beat-frequency cycles   76–77, 124, 128, 134–135, 144, 146, 156
black-body radiation   19, 31–34, 97
black holes   27, 183–184, 187
boundary layer   38, 97–101
bright spot   27–30, 55–58, 69, 85, 135, 145, 152

circularisation radius   24, 25, 58, 128, 132–133
Chandrasekhar limit   17, 51, 168, 182, 193
common envelope   46, 51, 163, 178

degeneracy   53, 162, 182
diagnostic diagram   190, 192
dips   103–104, 113, 116, 134, 153, 157
distances   39, 50
Doppler shifts   37–44, 102, 128, 141, 164, 189–192
Doppler tomography   43–44, 94–95, 115, 192

echo outbursts   90
eclipse   15, 16, 27–30, 55–57, 67, 68, 76, 112–113, 115, 137, 143
   mapping   39–40, 82, 105
Eddington limit   186
ellipsoidal modulation   21, 23, 29–30, 190
ER UMa stars   87–91
escape velocity   102, 111

flickering   30, 148, 151–153
Fourier transforms   127–128, 130–131, 146, 154, 159, 160
FU Ori stars   188

gravitational radiation   47, 52

Heisenberg's uncertainty principle   32
Hubble Space Telescope   15, 26, 50, 190

infrahumps   92–93, 107

Kepler's law/Keplerian velocity   15, 18, 19, 38, 48, 59, 97–98, 100, 129, 153

Lagrangian point (L_1)   20, 128–129
line profiles   32–34, 41–44, 101–103, 115

magnetic accretion
   accretion curtains   141–144
   blobby   112, 121–122, 132, 149, 158
   cyclotron emission   116–120, 122
   field strength   111, 120, 135, 149, 155
   flux freezing   109–110
   polecaps/columns   111, 120–122, 137, 140, 149, 158
   stream–field interaction   111–113, 120, 132–133
   threading region   112, 120, 137
   X-rays   111, 116, 122–123, 139, 141
magnetic braking   47–49, 53, 73, 178
magnetic turbulence   61
magnetosphere   110, 128, 138–140, 156
mass ratio, q 21, 29, 50, 77, 80–81, 91, 190–193
mass-transfer bursts   57, 87, 176–177
mass-transfer/accretion rate ($\dot{M}$)   30, 38, 46–54, 72, 88–89, 91, 97, 108, 148, 166–167, 169–170, 178–180

neutron stars   17, 104, 168, 182–185
nova eruptions   125, 161–168
novalike variables   73, 91, 108, 171, 176, 179
nuclear burning   16, 45, 161–163

$O - C$ diagram   84–85, 136, 175
optical depth   35–36, 43, 101
orbital hump   27–30, 69
orbital-period distribution   49–54, 81, 90–91, 108, 149, 180
    period gap   51–52, 178
oscillations (DNO, QPO)   153–158, 166
outburst
    amplitude   55, 58
    interval   55, 58, 88, 176
    profiles   69–71

Pauli exclusion principle   16, 162
P Cygni profiles   101–103, 107–108, 163–164
polarisation   117–118, 127
propellers   46, 145–149, 163

quasars   27, 187

red dwarf   17, 50
    magnetic activity   47, 125, 149, 173–176
    mass   17, 54, 193–194
    radius   19, 81, 193–194
    spectrum   33, 174
    temperature   17, 19
red giant   16, 45, 46, 167, 181
reflection effect   17, 190
resonance line   100, 102
Roche lobe   20–24, 45, 51, 81, 170, 193
rotational disturbance   42–43
RS CVn stars   181

siphons   97–99
spiral shocks   93–95, 135
standstills   73–74
star spots   74, 174
Stefan–Boltzmann law   19
stellar wind   47, 184

stream
    impact with disc   85, 103–104, 153
    overflowing disc   26, 104–105, 134–135, 157
    trajectory   21–25
superhumps   75–77, 82–85, 89, 91
    late   83–84
    negative   92–93, 107
    period excess   81–82, 92–93
    permanent   91
supernovae   167
superoutbursts   75–76, 86–87
SU UMa stars   75, 81, 91
S-wave   41–42
SW Sex stars   105–108, 135

tides and tidal locking   21, 25, 27, 28, 47, 77–79, 82, 181, 185
T Tauri stars   188

U Gem stars   75, 81, 91
UX UMa stars   73

viscosity (α)   56, 59–63, 71, 89, 95
VY Scl stars   105, 171–173, 180

white dwarf   15
    mass   17, 38, 193
    radius   19, 193
    spectrum   31, 105
    temperature   17, 19
Wien's law   32
winds   99–103, 106–108
W UMa stars   22, 181–182
WZ Sge stars   87–91, 160

X-ray binaries   104, 168, 182–185

Z Cam stars   73–74, 91, 171, 179
Zeeman splitting   118–120
ZZ Ceti stars   159

Made in United States
North Haven, CT
12 September 2023

41476774R00128